Financing Construction

Financing Construction

Cash flows and cash farming

Dr Russell Kenley

Professor of Construction

UNITEC, New Zealand

London and New York

First published 2003 by Spon Press
11 New Fetter Lane, London EC4P 4EE

Simultaneously published in the USA and Canada
by Spon Press
29 West 35th Street, New York, NY 10001

Spon Press is an imprint of the Taylor & Francis Group

Publisher's note:
This book was composed by the author using Corel VENTURA Publisher.

Printed and bound in Great Britain by Biddles Ltd, Guildford and King's Lynn.

British Library Cataloguing-in-Publication Data
A catalogue record for this book is available from the British Library.

Library of Congress Cataloging-in-Publication Data
A catalog record for this book has been requested.

ISBN 0–415–23207–4

Contents

Figures

Tables

Preface

There are times in one's life when certain comments or experiences shape one's attitudes, views and beliefs. This book arises from a few such significant comments and experiences, which together have formed in my mind particular beliefs about the financial structure of the construction industry.

The origin of this work reaches right back to my days as an undergraduate student, where I had the fortunate experience of receiving lectures on cash flow management from the regional accountant of a very large and successful construction company. He claimed that one reason to be a head contractor (and, he joked, the only reason for an accountant!) was to gain access to the cash flow stream, and the only people who could do that well were the head contractors. He further claimed that a well-managed business could achieve a minimum of 20% and up to 35% of annual turnover as retained funds, and that 20% was therefore available for reinvestment. This, he further noted, could be a lot of money for a firm with high turnover. As a gullible undergraduate, I took notes, I remembered the lessons, I believed him. I assembled, about those lessons, my view of the financial management of construction. Together with my supervisor, Owen Wilson, I went on to investigate construction project net cash flows as a PhD thesis, with those lessons forming a context, almost a *raison d'être*. Ten years later I became gradually aware that no one else seemed to have had those lessons. I started to challenge my beliefs.

However, both my research and practical experience have demonstrated the truth of his words. Through research, I discovered huge variation in individual net cash flow profiles and identified that conservative companies tended to have a different profile from modern strategic businesses. I observed, first hand, the pain and suffering experienced in the Australian construction industry during a period of severe contraction following the excesses of the late 1980s. I saw the cascading failures following economic downturns, I saw companies I knew to be insolvent continuing to trade—surviving entirely on their cash flow and desperately hoping for a windfall gain. I saw owners of small firms, with low profitability, managing to afford luxuries such as farms, yachts and expensive cars. I remained convinced.

Private sector work in 1990 was extremely rare in Victoria—at one stage it was rumoured that there was only one project available for tender in the entire city of Melbourne. Government work was supporting the sector, until a change of Premier led to a freeze on government expenditure on capital works (including continuing contracts) in economic reform measures. The effect was rapid and disastrous. Over one notable two-week period, on average at least one medium to large contractor appeared to fail every day. Others limped along for a while on existing projects, only to fail as those projects drew to a close. By this stage, I had worked for lengthy periods with three contractors, including small, medium and large operations. Following these events, only the largest survived (but even then, not the division I had worked in—now that does not sound good for my reputation, but I am sure it wasn't my fault). I also worked briefly for what has become one of the world's most successful property and construction firms, and there I saw a different approach to cash flow management. They not only survived the downturn, but thrived.

To me it was obvious that the events I was observing were completely explained by those undergraduate lessons. If firms were reinvesting funds (whether intentionally or unwittingly) then clearly they would experience problems as new work dried up. Cessation of inward cash flow would cause an inability to pay already committed outflow. They would be unable to meet their commitments unless they could release the cash locked up in their investments. This would lead to insolvency, maybe not suddenly, but almost inevitably. On the other hand, those who did it well would survive and exceed their competition. Cash flow management seemed to be the key to unlocking success in the construction industry.

For almost ten years, those years spent completing a PhD plus those that followed while working in industry, I was unaware that not everyone shared those beliefs. It did not matter at that time, because everything I observed (sadly, at that time, there was no shortage of strange and dramatic action in the industry due to the rapid changes in the economic environment) was explained by my belief structure. Gradually I became aware, however, that others didn't have the same explanations for events that I did. To me, their descriptions of events seemed simplistic and far too convenient. Company failures were blamed on clients not paying accounts, whereas I saw poor financial management by contractors as the most significant cause, an explanation that was neither palatable nor understood by the industry. I found myself in heated arguments and, as an academic, I realised that I was going to have to argue my case in a more structured way.

This work is the consequence of my attempt to support my beliefs and hopefully to convert some readers to my way of thinking. The reward is, I believe, to have a more financially aware and competitive industry and, in the end, with fewer company failures due to poor financial management.

This has been an immensely rewarding journey, in which the power of structured reasoned argument has generally prevailed over the prevailing rhetoric. In one-to-one debates the logic has held true. In industry seminars the support has been emphatic, if sometimes surprising. It remains to be seen what will be the effect of setting the argument out in this comprehensive form, where the clarity of logic can get lost amongst the academic jargon and apparently obscure modelling.

CONSTRUCTING THE ARGUMENT

I went in search of evidence to support my beliefs and started to think logically about the sequence of events that would lead to insolvency of a construction firm. Empirical evidence is hard to come by, so I have taken a largely reductionist approach.

These beliefs are best described by relating a common discussion which follows after a contractor is forced to close its doors. The scenario goes something like this.

- **Observer**: That contractor failed because they had a bad client who disputed and would not pay the final claims.
- **My response**: Hmm, interesting, I can see your point of view. Now tell me, When they failed they owed about $30 million on a turnover of about $150 million while the project they were constructing was worth about $8 million. So if the client was

responsible through delaying payment, there seems a bit of a gap in the explanation. Let us try an alternative explanation for their path to failure:

> The firm was successful and achieved a turnover of, say, $100 million. At some time in the past there was a project that went wrong. Not a major disaster but it is never good to lose money. However, the firm as a whole was very profitable, as evidenced by the very positive bank balance. Some investments are made in property and a few luxuries. Further down the track, a few more losses or a bit of a downturn in the market causes the positive cash position to dwindle. Things start getting tight. This is no real problem as the firm reduces margins to win work and maintain cash flow. After a while, other companies also reduce their margins in the tight conditions, and the financial position deteriorates. Even tighter margins are considered, possibly running to at cost or even buying work—always with the intention of winning back the margin by running the contracts aggressively. This strategy is successful and soon the company is winning more than its fair share of available work. (This is a clear danger sign indicating a distressed company.)
>
> Turnover reaches $150 million and resources are stretched, particularly on such low margins. Mistakes get made and litigation becomes common as project margins are protected. Unfortunately this leads to some delays in finishing work and getting paid. Clients are accused of causing problems for the firm by delaying payment. After struggling for a while (this phase could take years), the company runs into a client who is reluctant to pay (most likely citing problems with the quality of work). In desperation, the company asks its loyal sub-contractors to extend terms to 60 days—only temporarily, until they solve the problem of the late paying client. (This is another danger sign indicating a distressed company.)
>
> The company also tries to sell some investments, but this is hard and selling with prices at the bottom of the market seems unwise and not necessary—after all, it is only a temporary problem due to a late paying client. Winning more work will also help, so the company bids more aggressively in order to buy some work and certainly does not turn any work away, even from clients who might already have a reputation as poor payers.
>
> Suddenly, the whole system fails, due to nervous banks, sub-contractors, or simply weight of problems and an inability to move the disputes toward settlement. The ultimate cause might be something as simple as a delay in the projects caused by excessively wet weather. Whatever the cause, when the dust clears, $30 million is owed and the total cost to the supply chain may reach $50 million. Clients owe at least $8 million to the contractor at this time, but are unwilling to pay because of the cost of both completing and rectifying work.
>
> All parties agree that something has to be done about the problem in the industry of clients that do not pay their bills.

Most people from within the failed company would recognise how close this story is to the reality. They often tell me so. The question is, how does a firm which is only owed $8 million cause as much as $50 million damage? The answer is, rather than late paying clients, poor cash flow management. This view, supported by Davis (1991: 18; see page 9) is the opposite aspect to the undergraduate lecture. If a firm has 20% to 30% of turnover available for reinvestment (and making a loss is a form of reinvestment) then it is able to spend that amount before showing any signs of distress such as extending payment terms. On a turnover of $150 million, this amounts to $30 million to $45 million—sufficient to explain the final outcome.

This sequence of events is depressingly frequent in the industry, and it is surprising just how many failures fit this pattern. However, there is a bright side to the discussion. While there is a danger of being negative about the financial management in the construction industry, this is not fair as the majority of the industry operates well and is very profitable. Partly this is because of sound business management. Partly it is also due to successful companies understanding that, along with the risks, there is an extremely positive side to the strategic management of cash flow. It is this side which empowers a successful business to grow well beyond the extent which would be fundable through project margins. Understanding the underlying financial mechanisms is as much about understanding this positive side as it is about the negatives.

If theory is not your interest, then at least read Chapter 8. Here you will find a discussion about the strategic financial management of construction companies. Perhaps this will generate the same interest that I found all those years ago, and which introduced me to the theory—explored in the remaining chapters. I hope that in time there will be more research to support the conclusions in Chapter 8. I also hope that a new breed of energetic students and researchers will be inspired, as I was all those years ago, to shrug off the dominant thinking expressed in the literature from the ivory towers of academia to make their research real and relevant to the industry which it serves.

Russell Kenley
Auckland, New Zealand

Acknowledgements

Owen Wilson set me on this path. Owen is a quiet and dedicated quantity surveyor who came late to academia and discovered a love and enthusiasm for learning and above all research. Construction Management (Building as it was generally known twenty years ago) and Quantity Surveying, were not disciplines known for their founding on theory and research at that time. Owen grasped and was then imbued with love of questioning and exploring the underlying explanations for readily observable phenomena, often avoiding the most obvious in favour of a better solution. Owen must take much of the credit, if not the blame, for the ideas explored in this book. I owe an enormous debt to my former supervisor.

Professor John Scrivener taught me about integrity, dignity and the value of contemplation. My colleagues know I was a very poor learner. His support and encouragement, in the face of amorphous mediocrity, kept me in the battle. We will celebrate soon, but it is his turn to buy the wine. Redmans—Cabernet Sauvignon, please.

I owe an equal debt to my colleagues, those with whom I have been privileged to share the pleasure of discovery through research with rigour. These are my former and current PhD students. They have taught me so much about ways of thinking and interpreting, and refreshed that love for research. They have invigorated my enthusiasm, which is so easily dented in the modern academic institution where quantity is too often ranked higher than quality. These are truly exceptional people, and constitute our future. Those who have taught me most were Iain Jennings, now a management consultant; David Shipworth, now at Reading University; Michael Horman, now at Penn State University; and Kerry London, now at University of Newcastle.

We all owe so much to Dr Frank Bromilow, so much of the original work was assembled by him. It is time his contribution was lauded.

I acknowledge the financial support of the Faculty of Architecture, Building and Planning, at that great research university—the University of Melbourne.

I celebrate the support of UNITEC, tomorrow's university and now my home. They are leading the small and innovative nation of New Zealand in developing a research culture for the built environment. Their support, in particular from the leadership, has created the environment in which ideas can be developed and substantive works like this completed. Thanks to John Boon, Dean, Faculty of Architecture and Design, for creating the context, the academic freedom and the support. I would like also to thank my colleagues from the School of Building for creating an atmosphere conducive to hard work and productivity.

Of course I thank my family, particularly for refraining from asking why, even if they did insist on asking when.

Permissions

This work would not have been possible without the kind permission to use copyright material. This has been provided by:

Taylor and Francis, for permission to use material from *Construction Management and Economics*. http://www.tandf.co.uk

ASCE Publications, for permission to use material from the *Journal of the Construction Division.* http://www.pubs.asce.org/journals/jrns.html

CSIRO, Australia, for permission to use material from *Procedures for Reckoning the Performance of Building Contracts*. http://www.csiro.au/

Blackwell Publishing, for permission to use material from *Engineering, Construction and Architectural Management*. http://www.blackwellpublishers.co.uk/

Glasgow Caledonian University, for permission to use material from *Financial Management of Property and Construction*. http://jfmpc.gcal.ac.uk/

Extensive use has also been made of the published literature that forms the basis of the cash flow management in construction. This includes many diagrams and tables which have been reconstructed in this work from the base equations provided. Similarly much use has been made of information that has been reformatted to include new data, or consolidated to better follow the argument of this book. Every attempt has been made to acknowledge the source of the information used in this way.

Glossary

GROSS CASH FLOW TERMINOLOGY

Cash flow (generally)

General convention holds that the term 'cash flow' on its own indicates the 'inward cash flow' from the client to the contractor. Correspondingly, the term 'outward cash flow' is conventionally used for the payments out to subcontractors, suppliers and direct costs. This convention is specific to the construction industry. Accountants would generally hold that cash flow more broadly concerns flows of cash in both directions. To avoid confusion, the term 'net cash flow' will be used within this book for the combined cash flow and where specific differentiation is required, the term 'gross cash flow' will be used to separate the individual inward or outward cash flow streams from the resulting net cash flow. Cash flows are generally represented as a cumulative S curve. Cash flow is discussed in Chapter 3.

Commitment curve

A 'commitment curve' is similar to the 'work in progress', but instead of being the value of the physical work completed, it is the value of the work committed. This could form two measures. First, the value of contracts let. Secondly, the commitment to expenditure for work in progress. The latter is the usual use of the term. The difference between the commitment curve and work in progress results from the terms and conditions of the various contracts. There may be a difference in timing between work in progress and the commitment.

Curve envelope

The 'curve envelope' is a graphical region containing all the possible variations of a curve profile found by examining past data at a specified confidence level. Alternatively the curve envelope can be derived by the use of fast and slow tracking through a cost schedule.

Earned value

'Earned value' techniques monitor the value of work performed and the time in which it was performed, to assess the performance of the project. Earned value is a term which has been adopted in the project management community to describe integrated cost and schedule monitoring systems. However, here it is used to describe the general category of such performance systems. This includes a suite of integrated and remote monitoring systems discussed in Chapter 4.

Inward cash flow

The client-oriented flow of cash from the client to the contractor is termed the 'inward cash flow' and is interchangeable with 'cash inflow' and 'cash-in' and sometimes 'cash flow'. Cash generally flows in from the client in periodic payments called 'progress payments'. Inward cash flow from the client may therefore be seen as being a series of lump sums, usually at intervals of one month, with payments rarely received in between.

Outward cash flow

The flow of cash out to suppliers, subcontractors and direct costs is very different to the inward flow from the client. These payments follow the disparate contracts and agreements that exist between the contractor on one hand and subcontractors and contracted suppliers on the other, and also occur on an as-required basis as labour and materials are called up and used during the construction of the project. 'Outward cash flow' may be seen as an almost continuous (but variable) series of small lump sums, with a concentration about the end of the month.

Progress payments

The client-oriented flow of cash from the client to the contractor generally flows in from the client in periodic payments called 'progress payments'. Building contracts provide for such payments for two main reasons:

- To provide a mechanism whereby the contractor may recover money for work in progress, so that the contractor is not funding the project; and
- To restrict these payments to set periods (usually of one month) in order to reduce the amount of administration required by all parties.

Progress payments in procurement of major capital equipment are sometimes seen as a form of loan, to reduce the problem for the contractor of funding the work. The situation is similar in construction, however as work is installed, ownership transfers to the owner. This necessitates progress payments to reduce the risk to the contractor, and the concept of a loan ceases to be appropriate. For further details, refer to page 32.

Standard curve

A 'standard curve' or 'ideal curve' is produced by a cash flow model which takes the average of past data to produce a profile that is assumed to be standard within the industry for all projects, or subgroups of projects. There may be ideal curves sought for both gross cash flow and net cash flow. Standard curves are sought for their potential to allow forecasting of future cash flows. The value of standard curves has been questioned.

Work in progress

'Work in progress' (WIP) is the value of the work currently completed on the site. work in progress is sometimes considered an originating curve for inward and outward cash flow, as each are related back to the work in progress. The work in progress may be represented as cumulative work in progress and forms an S curve in the same way as inward and outward cash flow.

NET CASH FLOW TERMINOLOGY

Net cash flow generally

'Net cash flow' is the balance between inward cash flow and outward cash flow, and relates to individual projects. Projects usually commence with negative cash flow and then move to positive cash flow, eventually concluding at net cash flow equal to the project margin. Net flow is not linear, as would be suggested by the progressive accumulation of profit, but rather is highly variable and results from relative difference in the timing of inward and outward cash flow. Net cash flow is discussed in Chapter 6.

ORGANISATIONAL CASH FLOW TERMINOLOGY

Cash farming

'Cash farming' is a term used to describe the negative aspects of strategic or poor cash management practices. This is an emotive term, used because it reflects the unreasonable exploitation of the cash flow stream. The consequences of poor cash flow management are discussed in Chapter 8.

Organisational cash flow

'Organisational cash flow' is the cumulative addition across all projects of the individual project net cash flows. This includes all activities of the organisation, which may include other business units apart from construction. Organisational cash flow is discussed in Chapter 7.

Strategic organisational cash flow

Organisational cash flow should be managed strategically. This is strategic cash flow management. The company balances the cash demands from cash providing, low profit, activities with cash requiring, high profit activities. There are ways to do this well for long-term growth, as discussed in Chapter 8.

Working capital

Working capital has been defined as the funds required to cover the delay between payments and receipts (Davis 1991) during normal business activities. In construction, this is a critical issue as most contracts place demands on working capital at some stage and to varying extent. However, they also provide to the pool of working capital when they provide positive net cash flow.

1

Introduction

The construction industry is generally held to be fragmented, made up of a large number of relatively small firms which join together in temporary pseudo-organisations to undertake specific projects. At the crux of this organisation lies the head or management contractor, whose role, increasingly, is that of management rather than execution. This entity sits at a power locus, with all decisions and cash flow passing through their control. This is an enormously powerful position which carries a corresponding responsibility. The financial management of those firms in this position is what has created the industry as we know it today. This is an industry characterised by brilliantly successful financial management strategies which create wonderful projects, profitably, and generate wealth through competitive infrastructure. It is also an industry which, through woeful financial management strategies, has created loss, heartache, failed projects and cascading insolvency which has gone far beyond the direct influence of one contractor's projects.

This work aims to increase understanding of financial management issues, to help clients, contractors, subcontractors and suppliers to understand the system and to help prevent or minimise the consequences of contractor insolvency. The goal is to achieve greater understanding and support for those who manage the system well, as this will lead ultimately to a more competitive industry.

To achieve this aim, the book explores the theory behind cash flow management in construction, by introducing and comparing the various models developed by the research community to help to understand and simulate (or forecast) construction project cash flow. These models are set within a theoretical framework to assist in understanding the fundamental nature of cash flow management in construction.

Further, the issues of organisations and their strategic management are critical to understanding cash flow in practice. A lot of existing cash flow theory is based on assumptions about construction firms which simply do not hold up to practical examination. They ignore the actual behaviour of operators within the industry. So it is another aim to explore and highlight the way people behave when managing cash flow.

In part this is driven by a desire to help prevent businesses becoming insolvent due to poor organisational cash flow management. In part it is to achieve fair allocation of blame following such failures. But it is also about highlighting the opportunities which arise from strategic and careful management of cash flow.

While some of the views contained herein will be considered extreme by some, the most spectacular successes in our industry have come from mastering cash flow. Something which is so important to the industry needs to be debated and understood.

In summary, the aims are to:

- Understand the theory and the practical tools of cash flow.
- Improve the performance of industry through reduced insolvency.
- Improve the industry through improved competitive performance.

THE AIMS

Understanding the theory and the practical tools of cash flow

Cash flow has been analysed in components, but in reality they form a complete entity which should be understood holistically. It is partly through considering the components in isolation that the underlying mechanisms and behaviours become ignored. Cash flow, in construction, consists of the following layers:

- **Gross cash flow**—consisting of inward cash flow and outward cash flow. Founded on the relationship between time and cost, gross cash flow is closely allied with models for the time–cost relationship. Performance indicators are built on the relationship between time and cost performance, such as revising the forecast end-date and measuring the progressive value earned.
- **Net cash flow**—consisting of the balance between inward and outward cash flow. Founded on the interference between the component cash flow curves, and therefore not linked to the time and cost relationship. Performance measures are related to the time value of money.
- **Organisational cash flow**—consisting of the overlaying of the organisation's individual project net cash flows. Founded on a portfolio approach, the performance measures are the contribution to working capital and the minimum contribution. Secondary measures are the interest cost or income.
- **Strategic management of cash flow**—consisting of the policy and strategic framework for managing the cash flow for the entire organisation, including non-construction projects or investments. Performance measurement is the sustainable growth or well-being of the business.

Gross cash flow and net cash flow deal with cash flow on projects. Organisational cash flow and strategic management of cash flow deal with cash flow at the level of the organisation.

Improving the industry through reduced insolvency

Reducing contractor insolvency has immediate and easily understood benefits to the industry and to society. The knock-on effects through the supply chain mean that contractor failure affects many more than those immediately responsible. Most importantly, this also reduces the risk for clients of the industry.

Building economics, as a research discipline, has always had a steady trickle of papers resulting from the international research effort, but has never been particularly significant and rarely, for example, exceeding 10% of papers in the Journal *Construction Management and Economics*. It seems likely that such research is considered tangential to the real problems of production efficiency and service responsiveness.

Nowhere is the industry so well run that no financial failures occur; and the flow-on effects from such collapses resonate through the community. In 2001 in New Zealand, the industry iwas reeling from the effects of a series of major construction insolvencies, including the country's fourth largest builder. This has highlighted the damage that can be done through one company's mismanagement. Similar problems occurred in the early 1990s in Australia.

The gradual introduction of 'security of payment' legislation targeted specifically at the construction industry, originating from the UK, suggests that the industry has not solved its financial problems, with the result that the solution is being externally applied (through regulation). These solutions are in some ways bandage solutions, treating the symptoms but not the causes. It is only through improved research, education and debate that prevention may be achieved.

Improving the industry through improved competitive performance

In the UK, Sir John Egan produced the 'Egan report' *Rethinking Construction* (Egan *et al.*, 1988). This document ignited interest in rethinking the process of construction, based as it was on the premise that the industry as a whole is underachieving.

> It has low profitability and invests too little in capital, research and development and training. Too many of the industry's clients are dissatisfied with its overall performance (Egan *et al.*, 1988).

This has led to the development of intense research activity, primarily aimed at encouraging innovation in the industry. This focus, however desirable, has drawn attention away from research into the underlying functioning of the industry, and in particular its financial mechanisms and performance. However, an exploration of the economic contribution of the sector is important and, as we shall see in Chapter 8, essential to understanding the potential value of resources trapped and exploited within the industry.

Improved and strategic cash flow will improve the profitability of the industry. This is a good thing, as it has the potential to offer reduced costs for the client as well as providing better managed and better performing contractors. Strategic management of cash flow will be increasingly important for a firm to remain competitive in the market.

STRUCTURE OF BELIEF ABOUT CASH FLOWS

Denzin and Lincoln (1994) provide a basis for structuring thinking. While referring to the nature of paradigms, they outline a hierarchy in which paradigms[1], as basic belief systems, are based on *Ontological*, *Epistemological*, and *Methodological* questions:

> A paradigm may be viewed as a set of basic beliefs (or metaphysics) that deals with ultimates or first principles. It represents a world view that defines, for its holder, the nature of the 'world', the individual's place in it, and the range of possible relationships to that world and its parts, as, for example, cosmologies and

[1] A concept to be used with caution. The paradigm concept does not really help construction researchers. It is a high-level concept that tempts and distracts many research students. After all, who does not want to discover a new paradigm at some stage?

> theologies do. The beliefs are basic in the sense that they must be accepted simply on faith (however well argued); there is no way to establish their ultimate truthfulness.

The following order, quoted from Denzin and Lincoln (1994), reflects a logical (if not necessary) primacy:

> **1. The ontological question.** What is the form and nature of reality and, therefore, what is there that can be known about it? For example, if a 'real' world is assumed, then what can be known about it is 'how things really are' and 'how things really work'. Then only those questions that relate to matters of 'real' existence and 'real' action are admissible; other questions, such as those concerning matters of aesthetic or moral significance, fall outside the realm of legitimate scientific inquiry.
> **2. The epistemological question.** What is the nature of the relationship between the knower or would-be knower and what can be known? The answer that can be given to this question is constrained by the answer already given to the ontological question; that is, not just any relationship can now be postulated. So if, for example, a 'real' reality is assumed, then the posture of the knower must be one of objective detachment or value freedom in order to be able to discover 'how things really are' and 'how things really work'. (Conversely, assumption of an objectivist posture implies the existence of a 'real' world to be objective about.)
> **3. The methodological question.** How can the inquirer (would-be knower) go about finding out whatever he or she believes can be known? Again, the answer that can be given to this question is constrained by answers already given to the first two questions; that is, not just any methodology is appropriate. For example, a 'real' reality pursued by an 'objective' inquirer mandates control of possible confounding factors, whether the methods are qualitative (say, observational) or quantitative (say, analysis of covariance). (Conversely, selection of a manipulative methodology—the experiment, say—implies the ability to be objective and a real world to be objective about.) The methodological question cannot be reduced to a question of methods; methods must be fitted to a predetermined methodology.

These are useful, but by no means universal, categorisations of knowledge systems. They are extremely valuable in understanding the nature of cash flow research. Understanding the epistemology and the methodology of research, and ultimately the actual methods used within that context, enables the exploration of models to be undertaken.

Ontology

Much of the research into construction project cash flows assumes that they are something which may be modelled. It is assumed that they are real and have properties. This is a challengeable assumption, and it could equally be argued that cash flows are merely a collection of discrete events. However, in this work, cash flows are accepted as having reality as a collected entity made up of connected parts which exist through the existence of the project. Furthermore, it is assumed that they are additive and interrelated at the project level and also at the organisational level.

Epistemology

Epistemology is the philosophical theory of knowledge, of how we know what we know. It seeks to define knowledge, to distinguish its principal varieties, identify its sources, and establish its limits. The various sorts of knowledge seem to be reducible either to knowing-how (for example knowing how to speak Italian) or to knowing-that (for example knowing that Italy exists). Of these, knowledge-that is the prime concern of epistemologists, who differentiate between empirical and a priori knowledge. A priori knowledge is derived from its self evident axiomatic bases by deduction; empirical knowledge is derived from uninferred observation statements by induction. The usual sources of empirical knowledge are sense perception, while a priori knowledge is said to come from reason.

Accordingly, epistemology is generally characterized by a division between two competing schools of thought:

1. Rationalism (apriorism—or before experience, by deduction); and
2. Empiricism (from experience, by induction).

Rationalists insist that knowledge requires a direct insight or a demonstration, for which our faculty of reason is indispensable. The empiricists hold that all our knowledge must ultimately be derived, as it is in the sciences, from our sense experience.

So far the definition deals with how knowing-that is achieved, whether by deduction or induction. Epistemology is also concerned with the question of what it is for someone to know that something is so. The traditional assumption was that a person knows that P, if and only if (1) he believes that P, (2) P is true, and (3) he has good grounds for the belief. An alternative view is that knowledge is simply the ability, based on some non-accidental means, to provide the right answers.

Warnke (1987) casts further light on epistemology with an essay on Rorty's views:

> Rorty characterizes 'epistemologically-centered philosophy' as the search for a neutral foundation upon which to justify our beliefs and adjudicate between existing theories or interpretations. Since the seventeenth century this form of philosophy has defined knowledge as the correct depiction of reality, a mirror of the way things 'really' are with regard to the physical universe, the social world or the constituents of moral action. Accordingly the epistemological project has been that of showing the possibility of such accurate representation. The question it asks has been how the ideas and images within the human mind can be shown to correspond to an external world.

Epistemology is a term rarely but loosely used in the construction project cash flow literature. The bulk of cash flow research is empirical, being based on analysis of observed data. However, there is an increasing tendency to move beyond such analysis, to deduction, with the development of models for projects which do not arise directly from observation. Epistemology is not limited to rationalism and empiricism. Goldman introduces the division between the individual and grouped (social):

> Epistemology as I conceive it, divides into two parts: individual epistemology and social epistemology. Individual epistemology—at least primary individual epistemology—needs help from the cognitive sciences. Cognitive science tries to

> delineate the architecture of the human mind-brain, and an understanding of this architecture is essential for primary epistemology. Social epistemology needs help from various of the social sciences and humanities, which jointly provide models, facts, and insights into social systems of science, learning, and culture (Goldman, 1986: 1).

This second division of belief, borrowed here from the study of the human mind, introduces 'what is valid' in the understanding of cash flow models, which can be classified by methodologies following two epistemological categories:

- *Empirical* or *rational*;
- *Idiographic* or *nomothetic*.

These are neither mutually exclusive nor corresponding. Particularly, empirical studies may be idiographic or nomothetic, as may studies constructed on deduction. In this book, idiographic and nomothetic will be most often described as methodologies.

Methodology

Within construction cash flow research, the methodology is dictated by the epistemology. The basic forms taken are for empirical, post hoc, studies the (idiographic) analysis of individual project data, and the (nomothetic) analysis of grouped project data.

When forecasting or modelling future projects, another useful division emerges. This once again follows the idiographic–nomothetic division, and divides the research into:

- *Deterministic*: those methods which allow a *deterministic* model (usually from an average of past projects) to be held as a fixed model for future projects, from which forecast cash flows can be calculated.
- *Stochastic*: those methods which allow for the probability of variation in the forecast (sometimes for convenience from an average of past projects) by calculating future cash flows as an assembly of a trend and probabilistic variation about that trend.

Method

There are many methods used in cash flow research, whether empirical or rational, idiographic or nomothetic. They may however be broadly grouped, for convenience, into schedule-based or cost-profile methods. The former use the construction sequence, costed, to generate an originating cash flow profile. The latter utilises knowledge about the inherent properties of cash flow profiles to generate cash flow profiles for projects. These methods rely, to some extent, on the use of models to generate a representation of reality. Whether this be for schedule-based methods, which rely on the construction schedule as a model of the real project, or the cost-profile based methods, which rely on a model of the cash flow profile.

Modelling

A model is constructed to facilitate understanding and enhance prediction, and is an abstract description of the real world (Rubenstein, 1975). A model is not intended to be an exact description of the real world. In constructing a model, it is typical that simplifications and idealisations are used. If a model too faithfully represents all the features of the phenomena investigated, it may itself be too complex to be analysed (Singleton and Tyndall, 1974).

An 'empirical model' is one which examines a situation without going beyond the observational, measurement and recording level. A mathematical model which allows for convenient manipulation, and for the performance of the real subject to be approximated and predicted through the model can be alternately described as an 'analytic model'. Gates and Scarpa (1978) coined the word 'mathematized' to describe the step from observation to analysis.

Within cash flow research these models are generally mathematical. Shipworth summarised the relationship between the model and that which is being modelled, the modelling relation (Casti 1996), using mathematics as:

> The process of mathematical modelling requires defining what is to be modelled (setting the system boundary), selecting those observable features thought to determine the particular behavioural characteristic of the system being investigated (selecting observables) and encoding these observables into mathematical symbols (encoding). These symbols are then manipulated according to the rules of some branch of mathematics thought to mimic how those observables would behave in the system being investigated. Finally, these symbols are then decoded back into observable phenomena of the system (decoding). These decoded observables then constitute a prediction of the system's behaviour (Shipworth, 2000).

Either an empirical or a mathematical model may be used to generate a 'forecasting model', which is a means by which historic data may be utilised to predict future events. Many of the approximations and simplifications made in a model are justified on the grounds that forecasting is simplified, or made possible, by such action. However, a forecasting model should contain a measure of the confidence in its prediction.

STRUCTURE OF THE BOOK

This book has three main sections although they will not be identified as such. These are: context, consisting of this introduction and the economic context (Chapter 2); models at the construction project level, consisting of gross cash flow (Chapter 3), time related performance measurement (Chapter 4), the relationship between time and cost (Chapter 5) and the net cash flow on a project (Chapter 6); the final section relates to cash flow at the organisational level, consisting of organisational cash flow (Chapter 7) and the strategic management of cash flow (Chapter 8).

The intention is to develop the support through the theory of cash flow modelling at the project and then organisational level, to support the discussion of the strategic management of cash flow in Chapter 8.

It would be quite acceptable for a reader to commence reading with Chapter 8, and then to pursue those areas which provide support for the reader's needs in the earlier sections. However it is pursued, it will be seen that Chapter 8 is quite different in character from the rest of the book. This is because the discussion must move from the supporting theory into an area supported by reason and logic. This discussion is highly rational rather than empirical. This is not because data is not available, but rather because it is the interpretation of the data which is important. This is a topic where knowledge of the industry and the behaviour of people in the industry, becomes more important than mathematical models.

REFERENCES

Casti, J. (1992). *Reality Rules: I & II—Picturing the World in Mathematics*. New York, Wiley Interscience.

Davis, R. (1991). *Construction Insolvency*. London, Chancery Law Publishing.

Denzin, N. K. and Lincoln, Y. S. Eds (1994). *Handbook of Qualitative Research*. London, Sage Publications.

Egan, J. (1998). *Rethinking Construction*. London, Department of Trade and Industry.

Gates, M. and Scarpa, A. (1978). 'Pre-estimate cash flow analysis: Discussion'. *Journal of the Construction Division, American Society of Civil Engineers* **104**(CO1): 111–113.

Goldman, A. I. (1986). *Epistemology and Cognition*. Cambridge, Harvard University Press.

Rubenstein, M. F. (1975). *Patterns of Problem Solving*. New Jersey, Prentice-Hall.

Singleton, R. R. and Tyndall, W. F. (1974). *Games and Programs: Mathematics for Modelling*. San Francisco, W.H. Freeman and Co.

Shipworth, D. T. (2000). Fitness landscapes and the Precautionary Principle: The geometry of environmental risk, PhD Thesis, Melbourne, University of Melbourne.

Warnke, G. (1987). *Gadamer: Hermeneutics, Tradition and Reason*. San Francisco, Stanford University Press.

2

Economic context

INTRODUCTION

Recognition of the importance to the entire economy of the construction sector has increased in recent times. Attention has been directed toward making the industry more productive through improved efficiency and responsiveness, driven by innovation. Internationally, industry and government are driving a movement toward a research focus on innovation.

This new direction is to be commended. With the research will come new techniques, approaches, systems and solutions for the management of construction. However, within it lies a danger. The explosive investment in innovative applications and solutions often occurs without a detailed understanding of the underlying basics which drive the industry. There is a very real possibility that the benefits will fail through proposals being directed at the flames rather than at the seat of the fire.

One area of great importance is the financial management of construction and the way organisations manage their cash flow. Understanding this topic, the subject of this book, requires an economic overview. Management responds to external stimuli by changing behaviour. The changes in the economic environment over the last forty years have led to changes in the way the construction industry has managed its cash flow. It has gone from ad hoc management to strategic management, there have been times of deliberately holding back claims and periods of claiming early, and attention has waxed and waned as economic conditions have changed. These changes have come about through the impact of inflation and interest rates. In some ways it has been a wild ride, with periods of extreme fluctuation. Along the way many firms have become insolvent. But one thing is clear, construction industry managers can no longer ignore their strategic cash flow management, nor the prevailing economic conditions.

THE CONTRIBUTION OF THE CONSTRUCTION INDUSTRY

These days, any discussion of the economics of the construction industry must, it seems, discuss the contribution the industry makes to the economy and the corresponding potential for improvement.

The United Nations conducted a study in 1985 that attempted to identify the proportion of 'Gross Domestic Product' (GDP) in construction, comparing countries using the *real* value of final expenditure at international prices. They calculated the contribution of construction (residential, non-residential and other) to GDP. The resultant league table for a selection of countries is illustrated in Figure 2.1, with the UK (6%), USA (10%), Australia (11%) and New Zealand (11%) highlighted. These may differ from those obtained from national accounts at any given time, but they serve to illustrate that construction is a major contributor to the national effort.

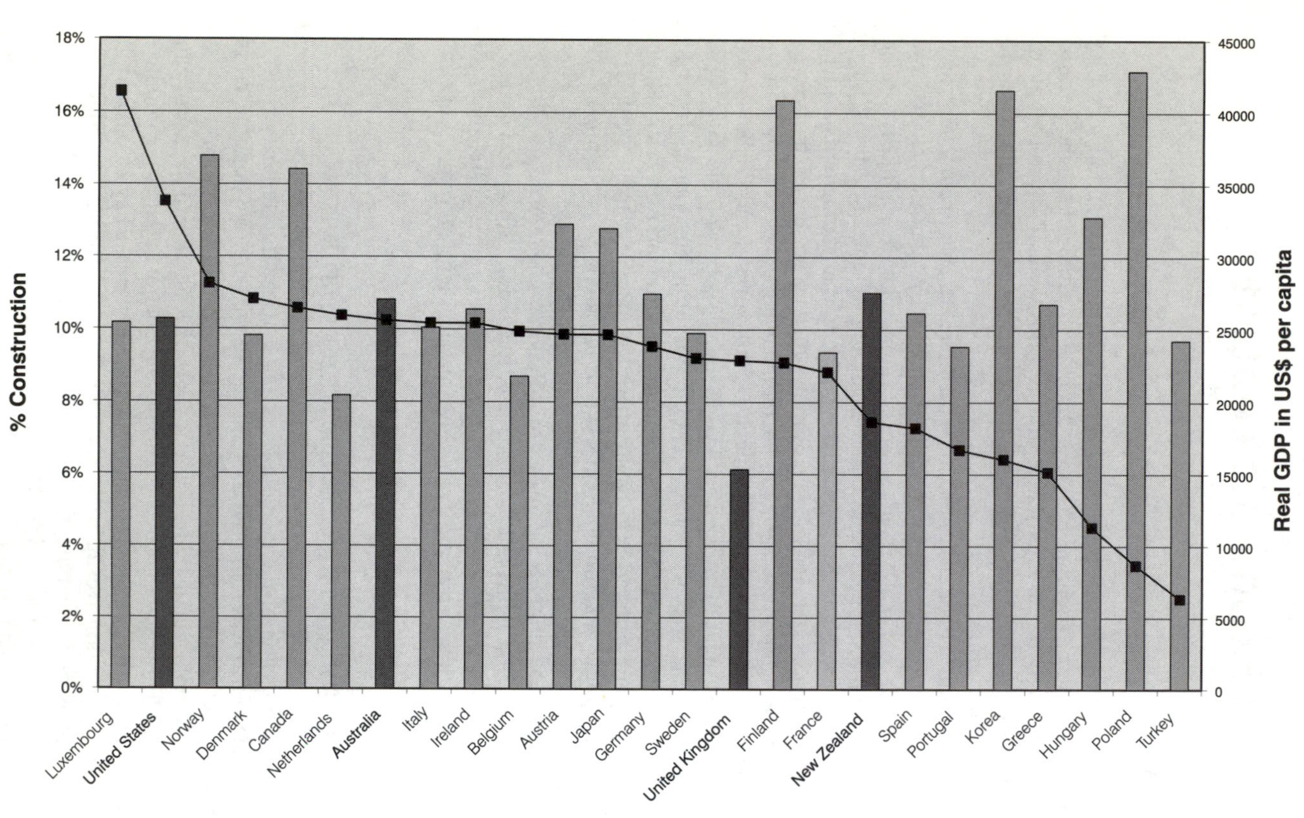

Figure 2.1 Real value of final expenditure on GDP at international prices and the percentage provided by construction sorted by decreasing GDP per capita

Further exploration of the extended United Nation's figures indicates that:

- Developing countries tend to have a greater concentration of effort in construction.
- Those countries with a very small contribution from construction tend to be those suffering severe economic difficulties.

Many researchers have placed the contribution from construction much higher than indicated by the UN figures, most often due to the inclusion of the flow-on effect of materials and supplies; the UK figure is supported by Hillebrandt (1985: 10). She excluded materials and supplies bought from other industries in her analysis. However, Hillebrandt differs in her interpretation of the share of GDP in other countries, here understating the figures significantly (3% to 10%) and arguing that while the lower figures belong to developing countries, there is a greater fluctuation in construction activity in developing countries than elsewhere.

The building and construction industry is an important contributor to the development and growth of a nation's economy, and there is a complex interaction between the two.

> The status of the building industry is commonly viewed as a prime indicator of the state of the nation's economy, especially in times of recession ('Employment and Housing', 1986).

This is because the building and construction industry is particularly sensitive to, and its reactions often precede, turn-arounds in the general economy.

The interaction is however two-way. There are direct relationships between the indicators in the economy and the performance of the industry. But there are also relationships between these indicators and the way the industry behaves. The following discussion looks at the movements in inflation (price escalation) and interest rates (cost of money) and draws a connection to changes in the financial management of the construction industry.

INFLATION AND INTEREST RATES

The building and construction industry precedes the economy as a whole, through a series of cycles, found by Lewis (1965; cited in Hutton, 1970), in his study of the British building industry, to be of 15 to 20 years' duration. While neither regular nor deterministic, Tan (1987) found six-year cycles in domestic construction (corresponding with cycles in fluctuations in) and twelve-year cycles in non-residential construction. These cycles are caused by variations in population, available credit, interest rates and shocks (unexpected events), and are therefore attributable to variation in demand. Similar cycles are interest ratesobserved in many countries, for example the USA (Tan, 1987).

Australian government influence in the building and construction industry up to the 1970s was principally through the control of credit. A government-sponsored credit squeeze affected the building industry in two ways; by lowering profit expectation among businessmen and thereby reducing investment; and by reducing the amount, and thereby raising the cost, of mortgage money for housing finance.

Though these factors remain current, the interaction of the economy and the building and construction industry has been complicated by a series of fundamental changes, including reversals, in the time value of money since 1973.

It is significant that Hutton (1970) did not include the topics of inflation and interest rates as significant factors in his assessment of the Australian building and construction industry. This may have been due to the low levels of each which had prevailed for the previous two decades; or that these factors, whilst present, were remarkably consistent.

Escalating inflation in industrialised countries in the early 1970s was a disturbing new feature in the post World War Two economic environment, for managers in corporations both large and small. Apart from making continued operations difficult for many building and construction firms, it forced closer attention to the basic skills of management. Better and more immediate information was clearly necessary (Green, 1977). Also required was better cost control and improved correlation between real costs and pricing mechanisms. Management discovered that a major task lay in training non-financial managers in the special impact upon their functions of inflation, and other time-related costs. Equally, in order to maintain control over operations, management shifted emphasis from long-term financial forecasts (which neglect the effects of the economic micro-climate) to short-term strategic plans (which take these effects into account).

Similarly, high levels of inflation caused problems for construction clients. Many long-term projects suffered excessively from price escalation. Clients became aware of the significance the timing of payments could have upon the final price of a building. Traditional building contract terms allow for reimbursement to the contractor for price escalation through what are called rise and fall agreements, which allow for payments to be indexed according to the time of payment. As the preparation of the index is periodic, timing can significantly affect the value of a payment to a contractor.

The last decade has seen a return to lower rates of both inflation and interest, reversing some of the drivers for management but not reversing their skills or knowledge which, once obtained, are available for use in the new context.

The impact of rising inflation

It can be seen from Figure 2.2 that inflation in Australia did not vary significantly from 3% p.a. during the period 1953–1970. A similar situation held for interest rates, as shown in Figure 2.6. The high levels of inflation of the immediate postwar period (reaching 20% in 1951) were probably considered an aberration arising from the extremely high demand for building materials following the war, whereas the deflation of 1962 (due to a credit squeeze with associated severe impact on the industry) had more immediate effects (Hutton, 1970). The period before 1970 was also one in which wages had high growth in real terms (due to claims for productivity gains) and prices were held down through restraint of the money supply (Hillsdon, 1977).

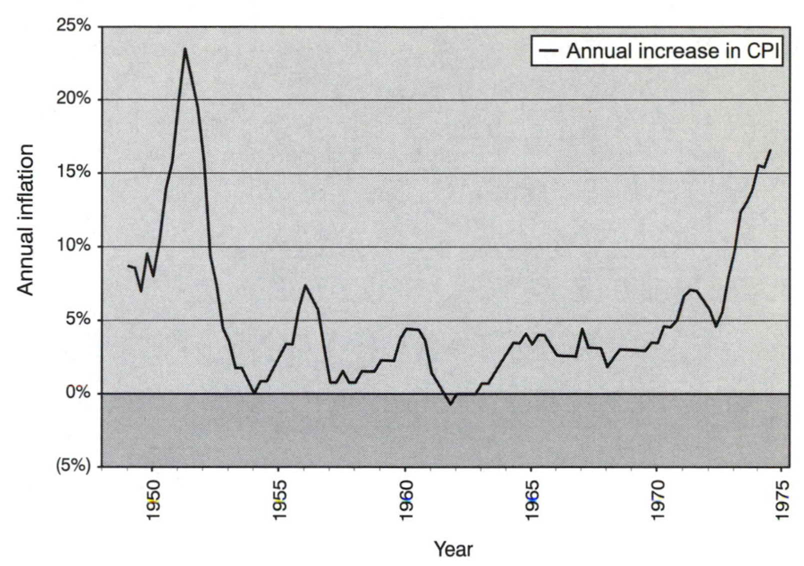

Figure 2.2 Australian postwar annual rates of inflation prior to 1973
(as measured by the consumer price index, all groups weighted average—ABS 6401.0)

The general picture for the construction industry prior to the 1970s was not good. Restraint of credit suppressed demand and there was a general shortage of building materials. General claims for productivity gains were being passed on to the industry while there were significant gains in productivity, due to improved methods of construction and to such innovations as ready-mixed concrete. These gains did not match those of other industries, which were undergoing automation and adopting more efficient work practices. Inflation by then must have seemed constantly low, given almost twenty years of constantly low escalation. The change to be seen thereafter must have been quite a shock for those involved in construction. Inflation from 1970 to 2000 is shown in Figure 2.3.

A combination of international and domestic forces caused a rise in Australia's inflation rate in the early 1970s, echoing the rise in inflation in Australia's major trading partners, as shown in Figure 2.4. The impact of the increase in inflation in 1973 on the building and construction industry was dramatic. Price growth in the industry was far in excess of the remainder of the economy, principally through the impact of wage rises on a labour-intensive industry. Inflation reached 17% inflation in 1975, compared with nearly 28% inflation in construction, as measured by the IPD for non-dwelling construction (ABS 5204.0). The Building Cost Index (BCI), which is published by the *Building Economist*, exceeded 26% for the same period, whereas the Commonwealth Wholesale Price Index for construction materials used in building other than housing (ABS 6407.0) only reached 20% in 1974. The former is based on a combination of the former and a wages index, and the two contrast the inflationary impact of wages at that time.

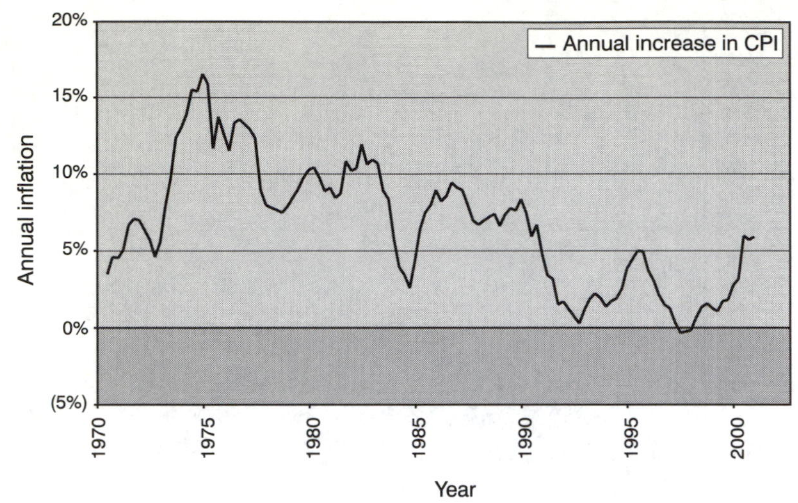

Figure 2.3 Australia's annual increase in CPI (rates of inflation) after 1970 (Australian CPI all groups)

The relative content of labour and materials varies from project to project. This necessitates that different ratios of labour and materials be used in the built-up cost index applied in the contract's rise and fall agreement. Such methods were introduced into standard contract forms, such as in 1977 the Australian Edition 5b standard form contract. During the periods of high inflation some large projects doubled in cost. Escalation clauses which were not tailored to the needs of a project caused financial problems for both contractors and clients. Many contracts had to be re-negotiated to avoid insolvency, which would not have been in the interest of either party. This was especially true of projects which had been entered before the need for rise and fall provisions had become apparent.

Inflation rates reduced from 1976 to the mid 1980s, but rarely to the levels of Australia's major trading partners (Figure 2.4) or down to the levels of the 1960s. The rates did stabilise between 5% and 10% which, although high, could be relied upon in the short-term to remain so for forecasting purposes. Extreme and changeable inflation rates create problems for management. Inflation has returned to lower levels of between 0% and 5% since the extremes of the mid 1970s and currently holds to low levels in most of the developed world. Management has now adopted practices which accept the existence of inflation. In fact, it may be argued, inflation is a boon for the construction industry, due to the related increase in capital value of projects.

Inflation has the effect of devaluing capital cost. Developers are able to utilise inflation to improve the viability of development projects by increasing the value of future income streams for inflation. This helps to explain why construction tends to boom in periods of high inflation, as capital for investment in future 'inflating' income streams is easy to obtain (the banks are willing to lend). Few other of the effects of inflation are favourable and when inflation is high, it requires new skills and practices to be developed.

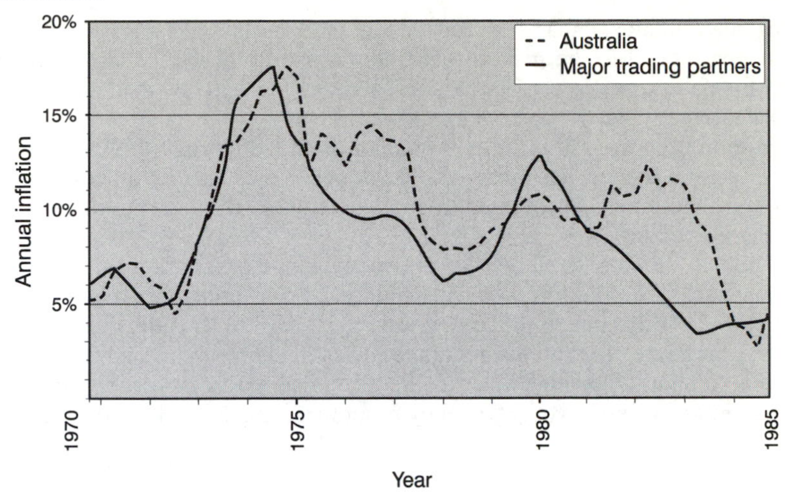

Figure 2.4 International comparison of inflation rates
(after Department of the Treasury, 1985)

Inflation caused management to concentrate on the correlation between real costs and pricing mechanisms. Emphasis was placed on the timing of claims for progress payments, both by the contractor and the client. Much research done at this time, and some previously unpublished work which emerged later (see Chapter 5), examined the time and cost relationships for building and construction projects. Models were produced by these authors which examined the sigmoid profile for cumulative claims (on average) in order to predict the staging and value of progress claims for a given project. These models will be examined in Chapter 3, and form the subsequent basis for cash flow modelling.

According to verbal evidence, client acceptance of these models was greater than contractor acceptance. This was reflected in client funding of the research effort, and of models aimed at the requirements of client bodies (for example Kennedy, *et al.*, 1970; Hudson, 1978; Bromilow and Davies, 1978; Peters, 1984), compared with later work which was funded by contractors (such as Khosrowshahi 1991, Kaka and Price 1993).

Client interest in gross cash flow forecasting arose due to a fear of price escalation, especially felt by large government and corporate bodies. They desired greater control over progress payments because they held the final responsibility for costs. There is further anecdotal evidence to suggest that contractor acceptance has been forced by clients who demanded forecasts of cash flows as part of the contract documents, and that contractors view such forecasts with a degree of scepticism.

Traditional contractual arrangements have rise and fall agreements which are open to abuse. The deliberate delay of a claim can be advantageous to a contractor if there is a large rise in the BCI expected in the following claim period. Allegations of such manipulations are by their nature difficult to prove, but it is easy to see that manipulations could be profitable in a period of high inflation, except when interest rates are higher, as experienced during the mid 1970s. It is often argued by the industry that it is still attractive to delay claims for rise and fall today, but such a claim ignores the counter impact of reduced overall inflation rates which prevail. Motivations related to the need to generate cash flow generally far outweigh the desire to delay payment for cost escalation, except in the most extreme circumstances.

With the relative fall in inflation rates from 1978 through to 2001, interest in gross cash flow models has changed. The concern of construction companies has swung from the impact of inflation to another time-related cost, the interest cost of financing operations.

There is therefore a relationship between the rate of inflation and the tendency to manipulate the timing of payments. Figure 2.5 illustrates this relationship, and indicates that a lower rate of inflation gives little reason to consider delaying claims, whereas a high rate of inflation encourages delayed claims to take advantage of cost escalation. It should be noted that this practice is quite rare except in conditions of very high inflation, but it is the tendency or motivation that is of interest here.

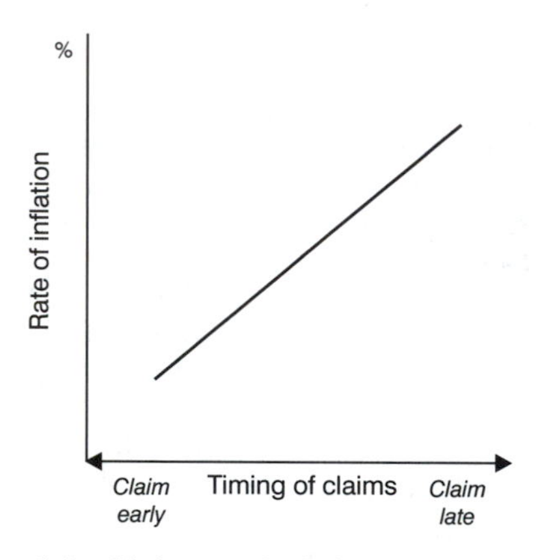

Figure 2.5 The relationship between the timing of claims and cost escalation

The impact of rising

Interest rates prior to 1973, as indicated by the rates for long-term government bonds, were about 5% (Figure 2.6). The building and construction industry was geared to low interest rates. There was common use of overdrafts to fund projects and little attention given to the timinginterest rates of payments. This situation was to change dramatically in the following decade.

At about the same time as an increase in the inflation rate was experienced, interest rates also began to rise. Figure 2.7 shows that long-term Australian government bonds increased from a stable 5% to around 10% from 1975 to 1980, and then again to approaching 15% from 1980 to 1990. It is only since 1990 that rates trended back through 10% to reach again the levels experienced before 1973. Similar results can be seen in other countries, for example the USA (Figure 2.8).

This general indicator, long-term rates, was echoed by the rates more applicable to the activities of building and construction industry; the Prime rate (large overdrafts) and the overnight money market rate (also known as the short-term money market, the Federal Fund rate (US) or the interbank market rate). These rates moved at the same time, as shown on Figure 2.9.

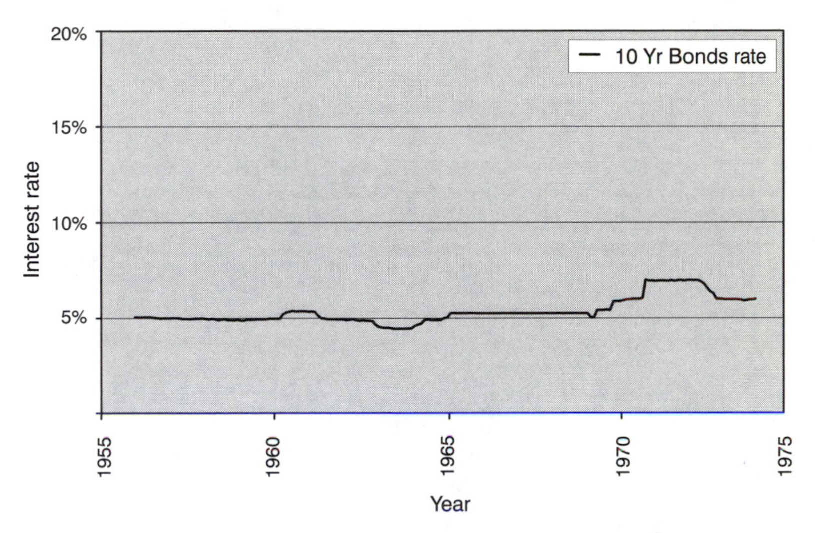

Figure 2.6 Australian interest rates prior to 1973
(ten-year Australian government bonds)

At times, the cost of maintaining an overdraft for a project in Australia has exceeded 20%. Similarly, the rates available on the short-term money market have also reached 20%. These rates are sufficient to require careful use of overdraft funding and to encourage exploitation of short-term investment strategies.

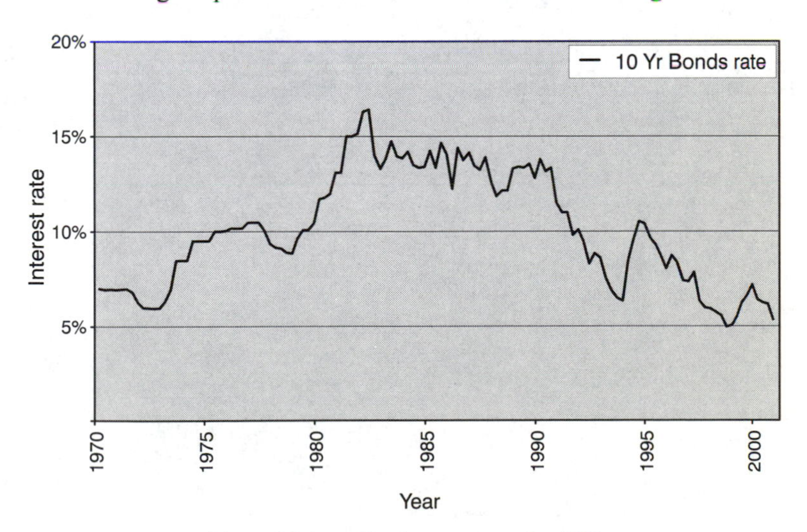

Figure 2.7 Australian interest rates after 1970
(ten-year Australian government bonds)

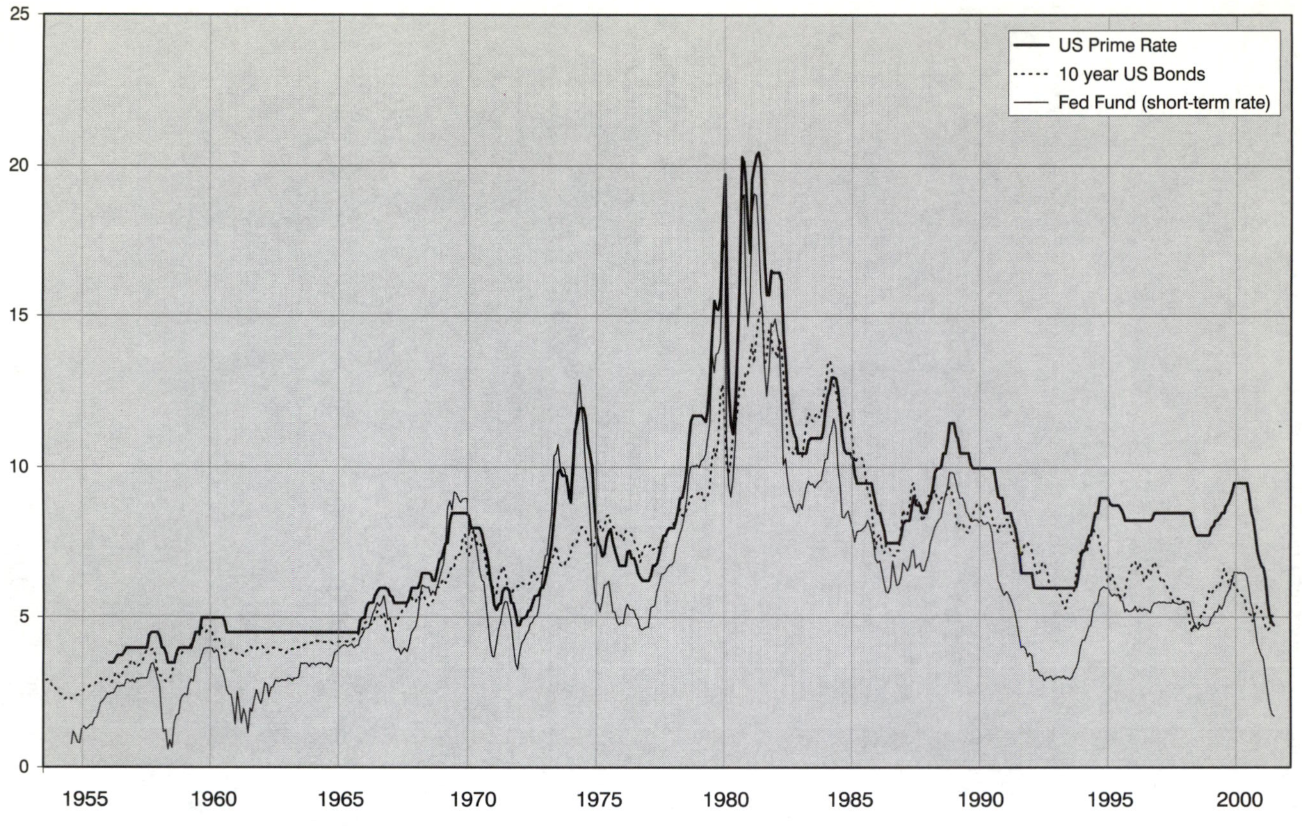

Figure 2.8 USA interest rates for the last fifty years, with ten-year bonds, the Prime rate and the Federal Fund rate (overnight or short-term) (USA Federal Reserve)

Interest rates moved upward at a time when managers were already being forced to concentrate on the time-related costs of construction arising from inflation. Management now had to cope with two fundamental shifts in the economic environment.

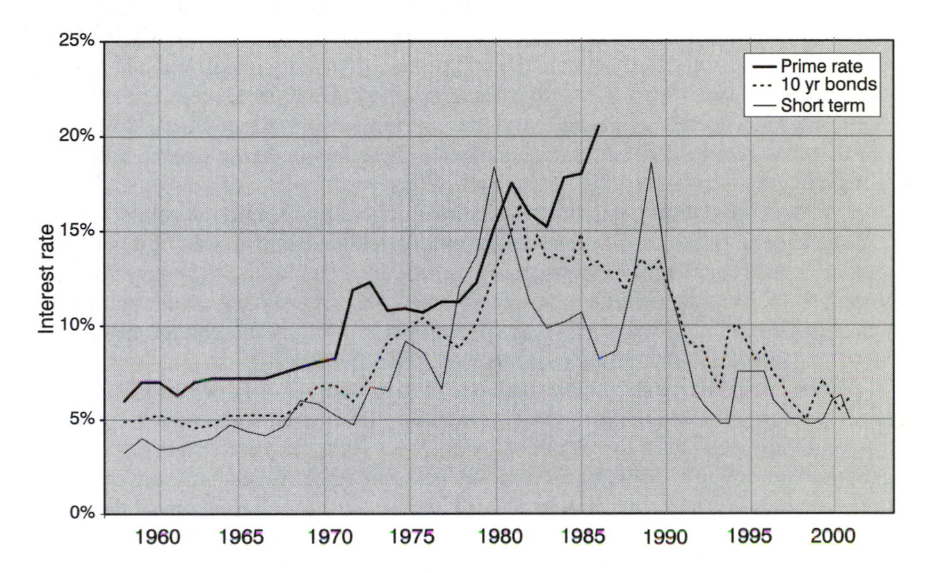

Figure 2.9 Australian interest rates for the last fifty years, with ten-year bonds, the Prime rate and the short-term rate (ABS Statistics, Ecowin)

Contractors began to claim early payment in order to reduce financing costs, and clients logically resisted such moves. For this reason much of the perceived advantage of playing the rise and fall game (a colloquialism for the delaying of payments in order to take advantage of inflated rates of payment in rise and fall agreements) was removed or exceeded by the interest costs incurred by the contractor. Further to this, clients felt an increased need for a forecast of cash flow over the project, not only to control their interest charges, but to allow them to take full advantage of other investments.

Inflation decreased after 1976, but for a while interest rates did not. Thus in the late 1970s attention swung from the apparently dynamic problem of inflation, to the more static problem of interest rates, and from gross cash flow models to net cash flow models with which the cost of finance can be managed. The period from 1973 to 1980 was a period of great change in the building and construction industry. According to Hutton (1970) the industry had previously been highly geared, with finance coming from overdrafts and creditors. Debt financing and under-capitalisation were major reasons for building company failure at that time (Hutton, 1970: 165). By 1980 many companies, although still highly geared, had ceased to operate at the limit of their credit. This was due to the costs of such finance, and because of the risks associated with high price escalation. For these reasons the overdraft ceased to be the primary means of financing building operations.

In the early 1980s there was a second fundamental shift in interest rate levels. For a time long-term rates moved from the new-found 10% level to a level in excess of 15%, but without the dramatic rise in inflation seen in the mid 1970s. The Australian prime rate rose from 12% to 17.5% in 1982, and after dropping to around 15%, exceeded 20% in December 1985. The rate for the short-term money market had gradually climbed to 10%, from 8% in 1979, and rose to nearly 15% early in 1982. It then entered a period during which it fluctuated wildly (daily levels reached as low as 4.6% in December 1983) before climbing to above 18% in December 1985. The period since has been one of steady decline returning eventually to the postwar levels. Apart from a rise in the period around 1995, there has been a steady decline right through to 2001.

What is most interesting about the movements in rates and the related changes in management behaviour, is that management realised that it was better to move from a model reliant on borrowings to one which reduced such reliance to reduce interest costs. Management has also seen the benefits for poorly geared organisations in avoiding financing projects through borrowings. In effect the change has become permanent, although the motivation has changed.

There is a relationship, therefore, between the cost of money (interest) and the tendency to manipulate the timing of payments. Figure 2.10 illustrates this relationship, and indicates that a lower interest rate gives little reason to consider claiming early (and paying late), whereas a high interest rate encourages early claims to take advantage of short-term investment rates. This is a similar relationship to the inverse relationship between interest rates and investments (Tan, 1987), where low interest rates equates to high investment.

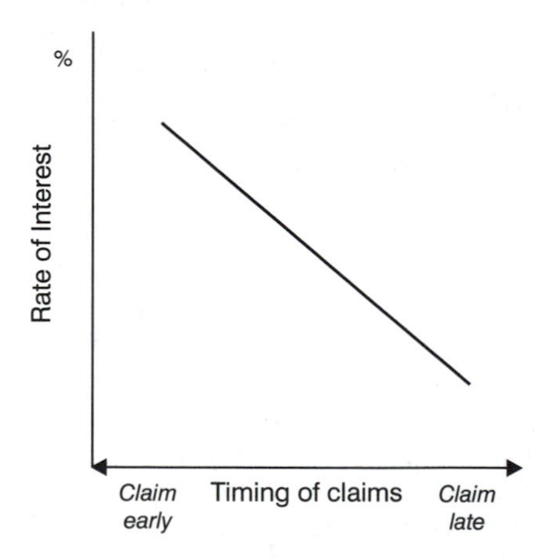

Figure 2.10 The relationship between the timing of claims and the cost of money

This practice is quite common, even in conditions of low interest cost, and anecdotal stories are quite colourful, with tales of clever practices such as putting all payments in a drawer at the due time and waiting for a phone call before paying: a simple technique which can delay payments sometimes by months. All this while at the same time having deals with banks to stay open late on Fridays to outwit clients who deliver cheques after 5 pm in the hope of delaying settlement until after a weekend. Such manipulations are discussed further in Chapter 8.

Real interest rates

Not only did interest rates increase to 1985, but they essentially destabilised. This change was complicated by a move from positive to negative 'real' interest rates in the early 1970s and then back to positive from the 1980s, as illustrated for Australia in Figure 2.11. In other words there was a move from interest rates above, to interest rates below, the corresponding inflation rate in the 1970s. This short period was unusual and had a significant impact on management behaviour.

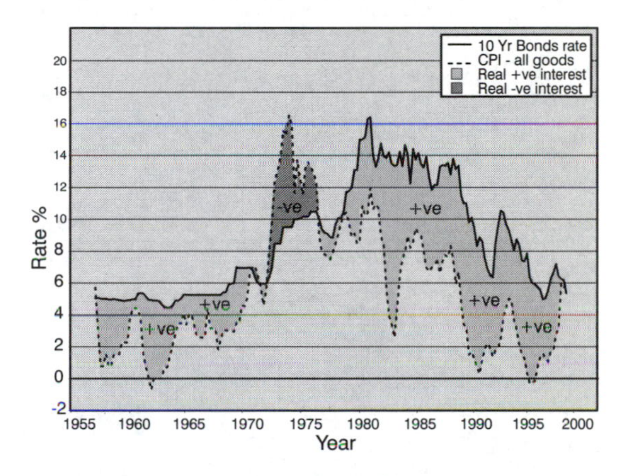

Figure 2.11 Effective 'real' interest Australian rates from 1970–2000

The equivalent charts for the UK (Figure 2.12), the USA (Figure 2.13) and New Zealand (Figure 2.14) show that similar patterns of real interest rates have applied in these countries. Akintoye and Skitmore, (1994) discuss the relationship between real interest rates and demand for construction and note that private sector demand for commercial building is sensitive to real interest rates.

There is a relationship between the real cost of money (real interest) and the motivation for management to manipulate the timing of payments. Figure 2.15 illustrates this relationship, and indicates that a positive real interest rate encourages claiming early (and paying late), whereas a negative real interest rate encourages late claims to take advantage of cost escalation. It should be noted that the ability to utilise relatively high interest rates, will increase a contractor's benefit from managing the timing of payments, by effectively increasing the real interest rate.

It is possible for a construction company to use trade credit, through tight management control, as a means for raising funds over that required for operations (see Chapters 7 and 8). This is counter to traditional building management, which tends to rely on borrowed funds to finance the difference between income and expenses.

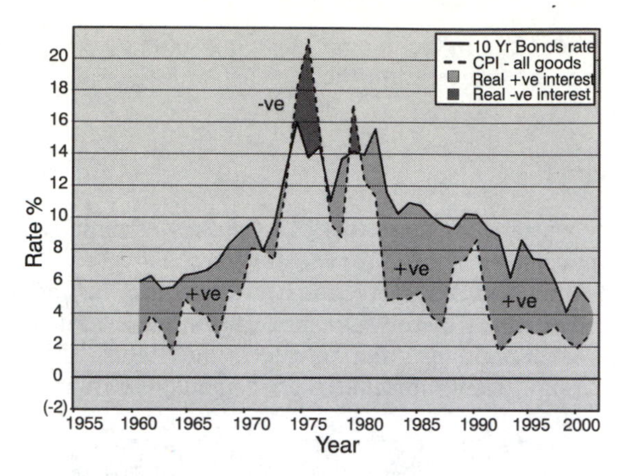

Figure 2.12 Effective 'real' UK interest rates from 1970–2000

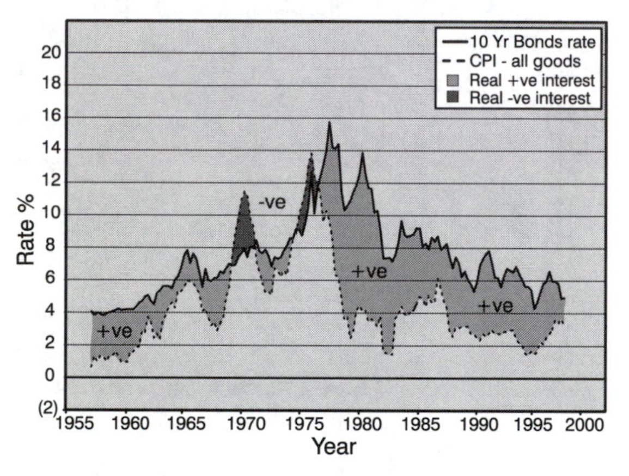

Figure 2.13 Effective 'real' USA interest rates from 1970–2000

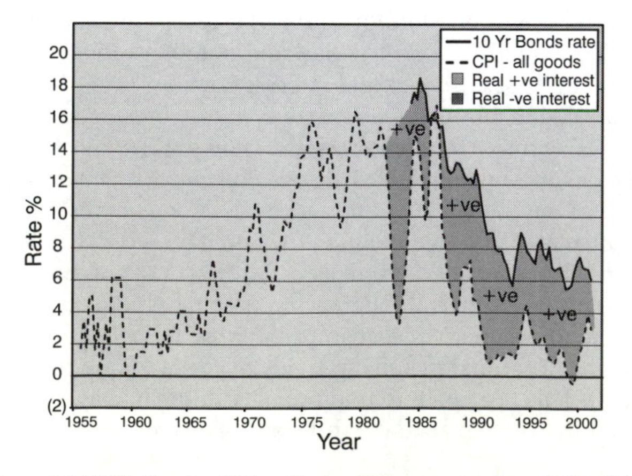

Figure 2.14 Effective 'real' New Zealand interest rates from 1970–2000

High and unpredictable interest rates make overdraft financing unattractive. Similarly, high returns on short-term funds make short-term investment attractive. The returns to be gained through investment may exceed those to be gained through operations, and so a major justification for running a building company is to raise funds through operations for investment.

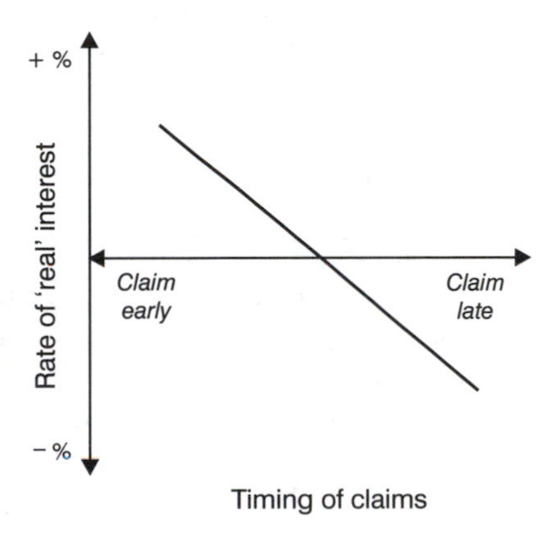

Figure 2.15 The relationship between the timing of claims and the real cost of money

While interest rates have not often been at consistently high levels, construction firms have, through experiencing high interest rates and also through experiencing negative real interest rates, become acutely aware of the balance between the money they are paying out, and the money they are receiving and the associated opportunity cost. They are thus prepared to expend much effort in achieving a positive working capital (sometimes called positive liquidity) on each project.

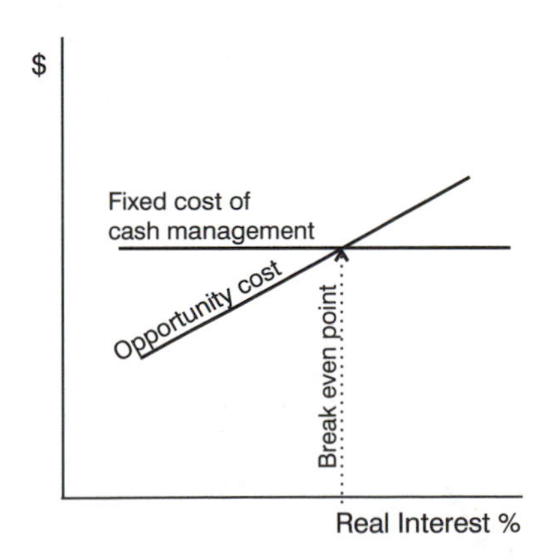

Figure 2.16 The break even point for returns from expenditure on cash management against increasing interest rates

There is another reason why contractors are prepared to spend resources to achieve positive working capital. The cost of the increased management control, in real dollars, is essentially the same as it would be in a period of low interest rates. However, the gains are relatively higher. Controls over funds would always have resulted in returns to business, but it is during periods of high interest rates that such controls become most profitable. Figure 2.16 illustrates the hypothetical break even point where the returns exceed cost at increasing interest rates.

It follows that in periods of positive real interest rates, particularly when as high as in the 1980s, building and construction firms perceived a need for tighter control over working capital generated or utilised by operations. This level of interest can be expected to differ from country to country as the relative real rates differ in each.

CONCLUSION

In 1973 most developed economies underwent a change which resulted in levels of inflation beyond the experience of managers within the construction industry. This change highlighted the need for management of the flow of cash on a project, especially that from the client to the contractor. Clients of the industry initiated, or supported the development of, construction project gross cash flow models. At about this time, and again in 1981, interest rates also increased. This resulted in construction companies becoming concerned with their net cash flows on projects, with a view to setting standards for management performance. Increased costs of finance, and profits to be made from the investment of surplus funds, encouraged contractors to devote resources to the control of working capital.

REFERENCES

Akintoye, A. and Skitmore, R. M. (1994). 'Models of UK private sector quarterly construction demand'. *Construction Management and Economics* **12**: 3–13

Bromilow, F. J. and Davies, V. F. (1978). 'Financial planning and control of large programmes of public works'. *Second International Symposium on Organisation and Management of Construction*, Haifa, Israel, Technion, Israel Institute of Technology.

Ecowin, Economics for Windows, www.ecowin.com

'Employment and Housing' (1986), *Rebuild*, Newsletter of the Division of Building Research, CSIRO, 11 (4) 2

Federal Reserve. *Statistical Release*, United States Federal Reserve Bank http://www.federalreserve.gov/releases/H15/data.htm

Green, J. (1977). *International Experiences in Managing Inflation*. New York, The Conference Board.

Hillebrandt, P. M. (1985). *Economic Theory and the Construction Industry*. London, Macmillan.

Hillsdon, B. (1977). *Hedging Against Inflation*. Canberra, Committee for Economic Development of Australia.

Hudson, K. W. (1978). 'DHSS expenditure forecasting method'. *Chartered Surveyor—Building and Quantity Surveying Quarterly* **5**(3): 42–45.

Hutton, J. (1970). *Building and Construction in Australia*. Melbourne, Institute of Applied Economic Research, University of Melbourne.

Kaka, A. P. and Price, A. D. F. (1993). 'Modelling standard cost commitment curves for contractors'. *Construction Management & Economics* **11**: 271–283.

Kennedy, W. B., Anson, M., Myers, K.A. and Clears, M. (1970). 'Client time and cost control with network analysis'. *The Building Economist* **9**(3): 82–92.

Khosrowshahi, F. (1991). 'Simulation of expenditure patterns of construction projects'. *Construction Management and Economics* **9**(2): 113–132.

Peters, G. (1984). 'Project cash forecasting in the client organisation'. *Project Management* **2**(3): 148–152.

Tan, W. (1987) 'GNP, interest rate and construction investment: empirical evidence from US data'. *Construction Management and Economics,* **5**: 185–193.

Note: International economic data was provided by Ecowin software. For further information consult their web page: www.ecowin.com
Ecowin consists of statistical software and an international time series database, providing on-line access to long-term data for many indicators in many countries, each provided by the national government statistician.

3

Gross cash flows

INTRODUCTION

This book serves a number of users, from the student to the practitioner. Therefore the discussion will range from the theoretical to the practical. In this chapter, we will examine the topic of construction project cash flows, being the specific omni-directional flow of cash to, or from, the contractor. Of interest are the practical issues of timing and size, the purpose and origin and the practical management of payments. The discussion will then move on to the need and role of cash flow forecasting. Once again this deals with the applied issues of cash flow management and management needs at key stages such as at tendering. This is of practical interest to both clients and contractors, and prepares the case for a more theoretical approach. The need for models for cash flow is further justified with a discussion of practical cash flow forecasting techniques, in particular the use of priced project schedules, before moving into the discussion of cash flow modelling. Here, the various schools of thought for modelling cash flows are discussed in an historical context, as introduction to the Kenley and Wilson (1986) Logit cash flow model, which underpins the analyses in the book. This approach is selected because it is designed to empower analysis of net cash flows, which is crucial to understanding the role of cash management in financing construction.

A substantial section of the chapter deals with a comparative analysis of the various cash flow models in the literature and their comparison with the Logit cash flow model. This analysis demonstrates that the Logit model is indeed capable of replicating almost any other curve profile developed thus far—and this demonstrates its capacity for forming the engine for sophisticated cash flow and net cash flow models.

The analysis includes an attempt to describe the development of the models over time and to chategorise the models by their epistemology and methodology—and relating these two as the models have shifted approach over time.

The discussion returns from the theoretical to the practical in the remaining section. Here the discussion revisits the understanding of the nature of cash flows in a practical sense but using the theory now available, thereby placing the models into the context of practical construction management. This is followed by a set of tools for practical and simple cash flow forecasting, including techniques for deriving parameters for specific project contexts. This section is intended to empower construction professionals to understand their projects and in particular to understand their forecasts. A worked example of calculating the parameters of the Logit model is proved, and these parameters are then used to calculate a forecast cash flow profile.

This is a large chapter, one which has attempted to do justice to the work of researchers working in this field. While an important topic in its own right, cash flow models are critical to the subsequent chapters on performance management and net and organisational cash flow. Thus, the attention to detail will hopefully draw reward in these later sections.

THE NATURE OF CONSTRUCTION PROJECT CASH FLOWS

Construction project cash flows are a sub-set of cash flow for the organisation. Construction project cash flow is the inflow of cash to the contractor from the client, and also the outflow of cash to the suppliers, sub-contractors and to direct costs (Figure 3.1).

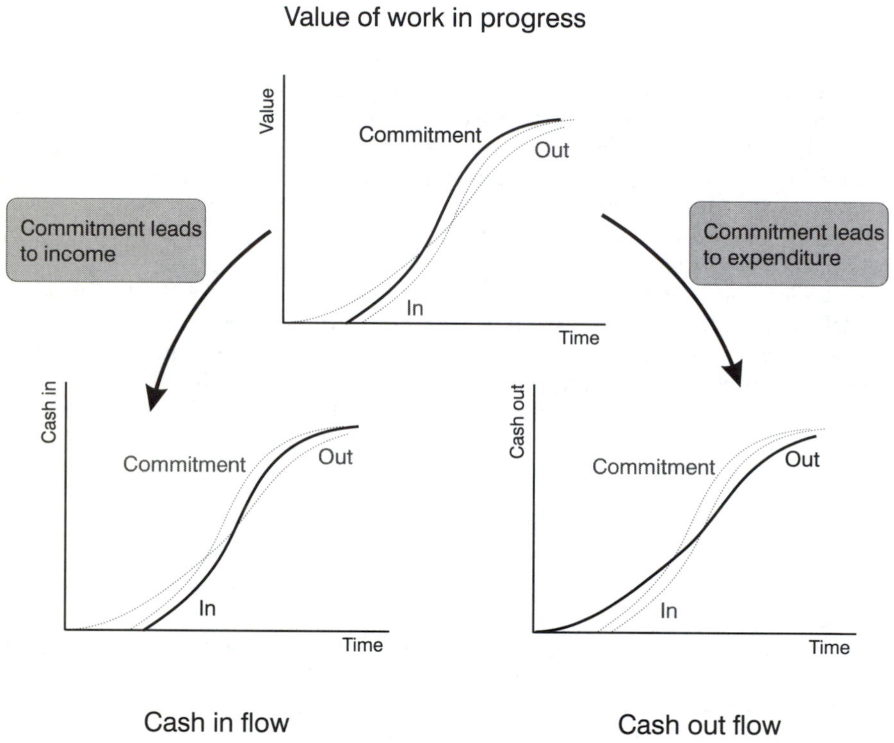

Value of work in progress

Value

Commitment

Out

In

Time

Commitment leads to income

Commitment leads to expenditure

Cash in

Commitment

Out

In

Time

Cash out

Commitment

Out

In

Time

Cash in flow

Cash out flow

Figure 3.1 Periodic: monthly, staged and turnkey project cash flows

Further terminology

Standard curve

A 'standard curve' or 'ideal curve' is produced by a model which takes the average of past data to produce a profile which is assumed to be standard within the industry for all projects, or subgroups of projects. There may be ideal curves sought for both gross cash flow and net cash flow. For example, the Bromilow model (the model constructed at the Division of Building Research, Commonwealth Scientific and Industrial Research Organisation (CSIRO); Kennedy *et al.*, 1970; Balkau, 1975; Bromilow, 1978; and Davies, 1978; Tucker and Rahilly, 1982, 1985) produces a standard cumulative outlay curve for government departments on public projects. Similarly Peer (1982) found a standard curve to express the flow of cash from the client to the contractor. The Department of Health and Social Security (DHSS) developed a model with standard curves for set project values (Hudson and Maunick, 1974) and Nazem (1968) suggests the use of an ideal curve for cash balances on all projects.

The reality of a standard curve has been questioned by many authors. Hardy (1970) stated:

> ...it is unlikely that a set of standard curves will provide a basis for cash flow forecasting for individual projects.

However, standard curves have achieved wide acceptance in the research community, particularly with regard to identifying greater sensitivity in categorisation in the hope of identifying more accuracy in ideal curves.

Curve envelope

The 'curve envelope' is a graphical region containing all the possible variations of a curve profile found by examining past data at a specified confidence level. Alternatively the curve envelope can be derived by the use of fast and slow tracking through a cost schedule.

Jepson (1969) and Kerr (1973) suggested finding the curve envelope by both these methods, and this work was later taken up by Singh and Phua (1984). Unfortunately the 50% confidence limits used by Singh and Phua indicate that their model is only appropriate for half of the projects analysed. Significant confidence levels (95%) would in many cases involve envelopes with proportions large enough to encompass almost all possible sigmoid profiles, and would thus be of doubtful value.

Inward cash flow

The client-oriented flow of cash from the client to the contractor generally flows in from the client in periodic payments called 'progress payments'. Building contracts generally provide for such payments for two reasons:

- To provide a mechanism whereby the contractor may recover money for work in progress, so that the contractor is not funding the project; and
- To restrict these payments to set periods (usually of one month) in order to reduce the amount of administration required by all parties.

Cash flow in from the client may therefore be seen as being a series of lump sums, usually at intervals of one month, with no payments received in between.

Outward cash flow

The flow of cash out to suppliers, subcontractors and direct costs is very different to the inward flow from the client. These payments follow the disparate contracts and agreements that exist between the contractor on one hand and subcontractors and contracted suppliers on the other, and also occur on an as required basis as labour and materials are called up and used during the construction of the project.

Payments may be made daily, such as purchases from local stores, weekly as wages or payments for goods on seven-day terms such as concrete or reinforcing steel, or monthly for subcontracts. At the end of the month many of the subcontractors and suppliers will be eligible for payment.

Outward cash flow may be seen as an almost continuous (but variable) series of small lump sums, with a concentration about the end of the month.

Types of periodic cash flows

Progress payments

Inward cash flow can take several forms depending on the nature of the contract. It was described above that cash flows generally occur in lump sums. The amounts of each payment are the periodic cash flows.

The purchase of major items of a capital nature causes problems for providers who necessarily suffer considerable expense before completing the capital item for delivery and therefore payment. The need to locate finance to enable the supplier to complete the work has resulted in the system of progress payments.

For capital works where the construction or assembly is undertaken at the supplier's premises, the supplier maintains ownership of the item until delivery. In this situation, highly structured forms of progress payment are required as the system of progress payments may be considered a loan.

> In their simplest form, progress payments may be viewed as a 'loan', a temporary interest free loan from a buyer (the prime contractor) to a seller (the subcontractor). They are based on costs incurred by the supplier in the performance of a specific order and are paid directly to the supplier as a stipulated and agreed to percentage of the total costs incurred (Fleming and Fleming, 1991: 2).

Within government jurisdictions there is a highly structured process for managing these payments which, due to the nature of the capital purchase, require this form of pre-payment or loan. Within the USA the Federal Acquisition Regulation (FAR) stipulates that:

> The supplier promises to 'pay back' the temporary progress payment loan by (1) making the contractual deliveries or by completing contractual line items, and (2) allocating some portion of the proceeds of the delivered unit price or completed line item values to liquidate the loan, based on the established subcontract unit price of the articles or services delivered.

This highly structured environment is actually quite different from that which pertains to construction due to the physical location of the works and the ownership of work in progress. In the event that the supplier fails, the client is in a relatively weak position and must therefore have tight regulations, restrictions and agreements in place.

Construction work is normally conducted on the client's land, or at least it is not on land controlled by the builder (this latter is normally under the category of development and is a separate issue). As such, the builder is in a weakened position as far as cost recovery in the event of, say, failure of the client. Thus there is a need for the builder to recover cost progressively and it should not be considered as a loan. In construction the relationship is more akin to a service (constructing a facility on the client's land) than the provision of a capital item (delivery of a completed building).

> Every construction operation, with the exception of offshore work, is concerned with a building or a structure created by fixing materials to land. This simple fact has a profound effect on the way the industry functions.
>
> As each item is fixed to the structure, it loses its separate identity and becomes part of the land. Although the contractor gives possession of the structure on practical completion, and, in a sense, delivers the works, this cannot be compared with delivery of goods by a manufacturer. This is because ownership of the goods had been lost to the employer long beforehand on their incorporation into the building (Davis, 1991).

There are payment systems which replicate the delivery of a capital item, such as Turnkey where payment occurs at handover only, but staged payment systems, such as monthly progress payments and staged progress payments, are more common.

Monthly cash flow

For most projects, the inward periodic cash flows are monthly. A table of monthly periodic cash flows therefore illustrates the actual amounts paid at the end of the month (Table 3.1: Monthly column).

Staged cash flow

It is useful to compare commercial/industrial building with domestic building. Generally for commercial/industrial contracts, a contractor is entitled to claim for payment at the end of every month, whereas a house builder is may be entitled only to payments at the completion of stages in the progress of the works—such as at lock-up (Table 3.1: Staged column).

Table 3.1 Comparative table of periodic cash flows

Stage	Period	Monthly	Staged	Turnkey
	Jan	$0		
	Feb	$382,603		
	Mar	$1,350,726		
Stage 1	Apr	$2,557,667	$4,290,996	
	May	$2,543,151		
	Jun	$1,794,395		
Stage 2	Jul	$894,551	$5,232,096	
	Aug	$377,534		
	Sep	$97,330		
Completion	Oct	$2,044	$476,908	$10,000,000

Turnkey cash flow

There are other methods for payment by the client to the contractor, particularly with modern contracts. One method that is becoming more common may be described as

Turnkey. Under such contracts, a single payment only is provided for at the conclusion of the project. This requires the builder to finance the project during construction, and involves a significant shift of risk to the contractor (Table 3.1: Turnkey column).

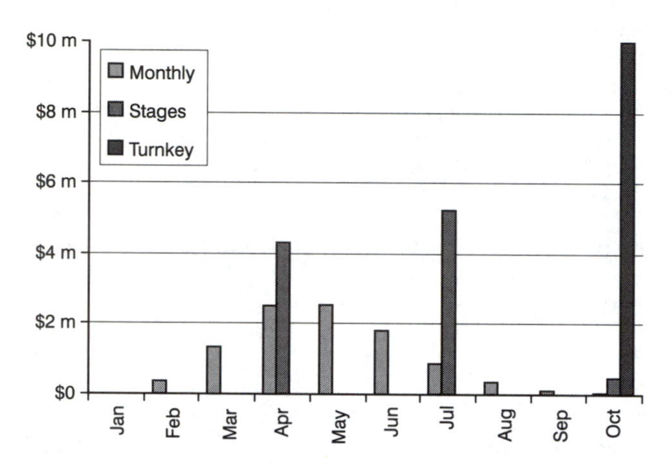

Figure 3.2 Periodic payments for monthly, staged and turnkey project cash flows

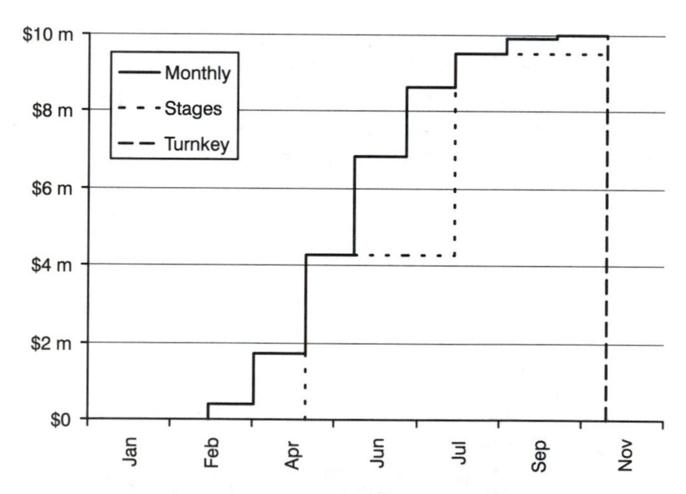

Figure 3.3 Cumulative payments for monthly, staged and turnkey project cash flows

Figure 3.2 illustrates the periodic payments for monthly, staged and turnkey projects. For the analysis in this chapter we will be primarily concerned with project periodic progress payments which are claimable monthly, as most commercial projects have contract terms which involve monthly payments.

Figure 3.3 illustrates the cumulative cash flow for each of these cash flow types, with the nature of the flow appearing as a series of steps, reflecting the lump-sum nature of the payments.

Cumulative cash flows

Cash flows may be represented cumulatively rather than as periodic payments. The cumulative representation is most common in the building industry, and is often confused with periodic representation.

Table 3.2 compares periodic and cumulative cash flows, which are best illustrated by graphing the cumulative cash flows as payments with the cumulative cash flow as a flowing line (Figure 3.4).

Table 3.2 Periodic and cumulative cash flow

Period	Periodic	Cumulative
Jan	$0	$0
Feb	$382,603	$382,603
Mar	$1,350,726	$1,733,330
Apr	$2,557,667	$4,290,996
May	$2,543,151	$6,834,147
Jun	$1,794,395	$8,628,542
Jul	$894,551	$9,523,092
Aug	$377,534	$9,900,627
Sep	$97,330	$9,997,956
Oct	$2,044	$10,000,000

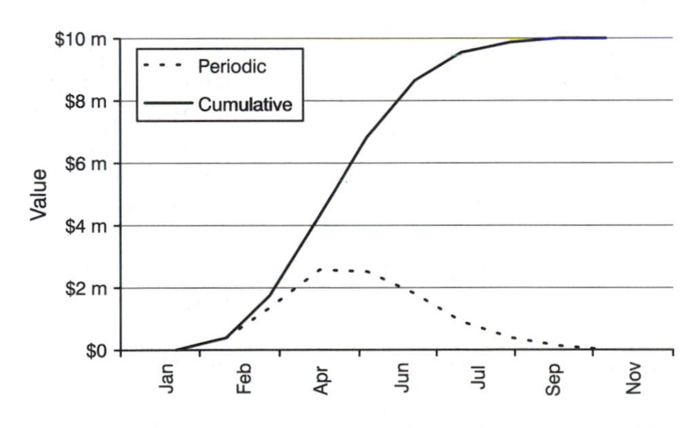

Figure 3.4 Periodic and cumulative cash flow

Tabular representation of cash flows

The most common form of expressing cash flows between interested parties, particularly clients, is the tabular form as illustrated in the previous section. The tabular form typically represents a cash flow amount at given time periods. This may be used, for example, to indicate a series of progress payments that are forecast for a project.

For example, tender documents often call for a forecast project cash flow to be provided within the tender. In such circumstances, it would be appropriate to provide a table of cash flows as above. This may be provided in either the periodic or the cumulative form.

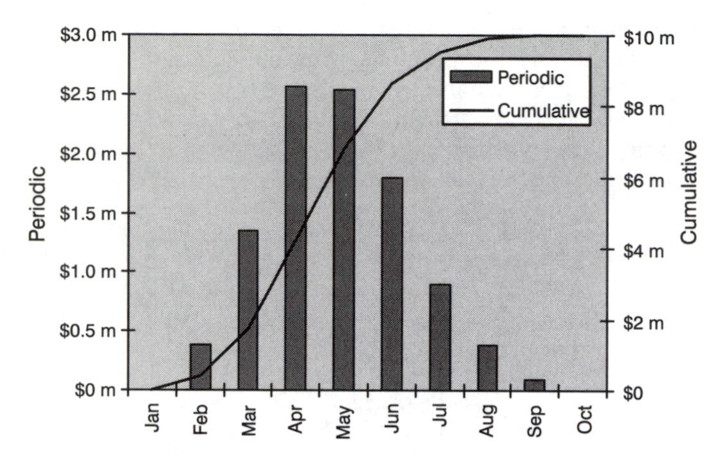

Figure 3.5 Alternative cash flow representation

Graphical representation of cash flows

Instead of using the tabular form of representation of cash flow, graphical presentation may be used. This form is particularly useful for visual comparison of project data and for trend analysis. The graphical presentation may be either periodic or cumulative. The cumulative form is, by convention, the most frequently used.

It can be seen clearly in Figure 3.4 that the cumulative curve approximates an elongated 'S' shape, a sigmoid curve. For this reason it is commonly referred to as the S curve. The slope of an S curve may be seen to be proportional to the periodic cash flow height. In other words, the greater the periodic cash flows, the greater the rate of increase of cumulative cash flow.

Figure 3.5 illustrates cash flow with the periodic payments indicated by a vertical bar, and the cumulative cash flow illustrated as a line. The scales are adjusted to make the patterns clearer.

The cumulative distribution, or S curve, is commonly used as a rapid means of forecasting the cumulative cash flow at a given period. For example, in Figure 3.4, it would be possible to draw a vertical line at a specific date and read off the value of the intersection of this line with the curve. This method is illustrated in Figure 3.6, which indicates a value of approximately $6.8 million at May 1994, and compares with $6,834,147 shown in Table 3.2.

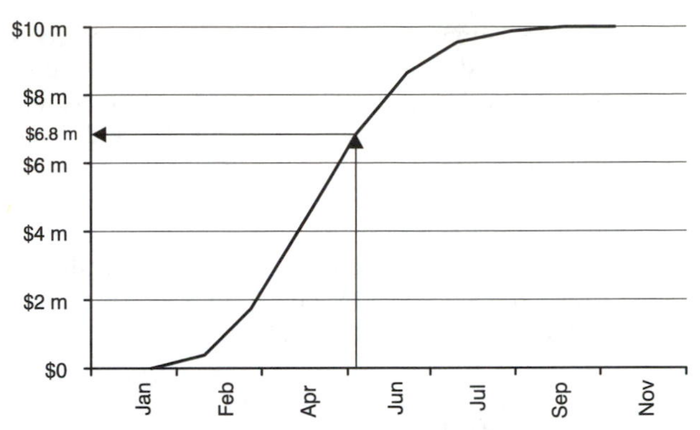

Figure 3.6 Reading the cumulative value from the chart

THE NEED FOR AND ROLE OF FORECASTING CASH FLOWS

The contractor's need for a cash flow forecast

There are many reasons for understanding and modelling cash flows. One common justification is the frequency of contractor failure.

> The success of a company depends on the ability of the management to foresee and prepare for the future. A contractor can use cash flow forecasting techniques to improve her/his financial position and remove the risk of bankruptcy which threatens so many construction companies (Evans and Kaka, 1999).

While this is a noble concern, it is not certain that monitoring gross cash flows will really 'remove the risk of bankruptcy'; this is more the arena of the net cash flow models to be outlined later. However, it is certainly true that a company which manages net cash flow is likely to begin by managing the gross cash flows for projects.

The client's need for a cash flow forecast

The client has a very real need for a cash flow forecast, and accordingly the contractor has a requirement to provide a reasonable forecast when requested in the contract.

Contractors become used to dealing with large sums of money. A progress claim for a million dollars or more is not unusual. In the demonstration cash flow above, we are calling for progress payments of up to $2.5 million. A contract administrator prepares the claim based on the work completed to date, claims and then awaits the payment. If the payment is late or incomplete then the panic button is pressed. The contract is read, the financial position of the client is checked, reminder letters are sent, etc. All this seems quite reasonable to the contractor. But what is the client's position?

Put yourself in the position of a client who had been expecting a claim for $2 million according to the cash flow forecast but instead at the end of the month received a claim for $2 million. The alternatives would be to pay $2 million at the due date, or to wait another week or so until they have found the additional $0.5 million. The next month, if the same was to occur, they would have to find an extra $1 million. Potentially when the next payment was due the contractor could be short paid $1 million which could jeopardise the financial viability of the contractor. If the client had sought a cash flow forecast as part of the contract, and this had called for the payment of $2 million then it might be argued that the contractor should carry the burden of an additional $1 million. A dispute would likely arise.

Reasons for a cash flow forecast

The client requires an accurate forecast of cash flow for very good financial reasons, including both direct and indirect costs. As such the forecast may form part of the clients investment planning.

Direct costs

The client may incur direct costs as a result of an inaccurate cash flow forecast. An accurate forecast would therefore be required:

- **So that funds are available**—The client must make sure that funds are available to settle accounts. They may have funds tied up in long-term investments, term deposits, property, shares, or nonliquid assets. They would then have to realise these assets in order to settle progress claims. Early realisation may be difficult, or may involve loss. For example, early realisation under duress of real estate is most likely to result in a sale at a discount (a 'fire sale').
- **So that interest is not lost**—The client will wish to maximise returns through investment. Investments may have to be converted from high-yield but long-term investments to low-yield but short-term investments, so those claims may be paid. Interest earnings will have been lost.

Of the direct costs, the former is perhaps the more crucial. If funds were tied up in term deposits with penalties for early withdrawal, it is easy to see the cost associated with an early claim by the contractor. For example in our $0.5 million higher than expected claim, this amount might be invested at 10% maturing next month—with no interest payable for early withdrawal. This would result in a loss of $4,167 interest, equivalent to employing another staff member in that month. In contrast, if a property valued at $0.5 million had to be sold early at a 5% discount, this might be a cost of $25,000.

Opportunity cost of funds

The second reason is perhaps not as costly, but represents a financial burden to a client with otherwise good financial management. Having extracted funds from high-yield investments, it would be extremely annoying to have these sit around in a low interest bank account for an extra month. In our example this might result in a loss of 7.5% interest or $3,125 per month. This becomes more significant in times of high interest rates, such as during the 1980s.

An investment plan

The cash flow forecast may be used by the client as a plan to assist in the management of investments. Such a plan would take a long-term approach to converting investments to funds for capital works payment.

Recovery of costs

A client incurring costs such as those above may seek to recover these from the contractor. If it could be established that the cash flow forecast was used as a financial plan by the client, then the client may be able to substantiate a claim against the contractor for varying from the plan.

Claiming to schedule

Many contractors, recognising these problems, set the cash flow as a budget and work to that budget—even if that means slowing the progress of works. It must be noted that this does not mean that the contractor should invoice according to the schedule without due regard to work in progress. This might be attractive to a client where progress is ahead of schedule, but certainly would not be attractive where progress is behind. Similarly, the reverse applies for the contractor who would have to bear the financing cost of delay in payment. A key role of the Quantity Surveyor is to monitor claims against the work in progress to ensure payments claimed have been earned.

Self fulfilling forecast

It can be argued that a cash flow forecast may in fact become a self-fulfilling prophecy. If the contractor cannot afford to work ahead of the claims schedule, and the client won't accept claiming behind schedule, then a contractor may choose to regulate progress to fit the target schedule. This is an important reason for the attraction to contractors of 'design and construct' contracts with schedules of payment. The ability for a contractor to design cash flows to generate significant funds is discussed in Chapter 8.

CASH FLOW AS A COST MANAGEMENT TOOL

The cash flow forecast and its indicated rate of expenditure may be used as a cost management tool by both the client and the contractor. The management of projects by monitoring project cash flow is an important function. This will be discussed in more detail in the Chapter 4.

By the client

The client can monitor progress claims against the schedule—and seek explanation. For example, if expenditure falls behind the forecast, this may indicate that progress has slowed on the site and that problems have been encountered. It may also indicate that the contractor is under-claiming for some reason. Similarly, if the claims are ahead of the schedule, this may indicate that the contractor is over-claiming. The client should in these circumstances be even more diligent in assessing the progress claims against actual value of work done.

By the contractor

The contractor can similarly monitor progress against the cash flow and use this as an early indication of problems. This may identify problems in the management of the project but may also identify cost over-runs. If progress on the project is to schedule, but costs are higher than planned, then there may be a problem. Early identification of the cost over-runs may assist in solving the problem before too much damage is done.

PRACTICAL CASH FLOW FORECASTING

The study of construction project cash flows has become increasingly popular over the last thirty years. Several approaches to the analysis have been used and early models may be characterised as nomothetic, in that they attempted to discover general laws and principles across categorised or non-categorised groups of construction projects, with the purpose of a priori prediction of cash flows. In contrast, recent studies have tended toward an idiographic methodology; the search for specific laws pertaining to individual projects.

Practical cash flow forecasting is an idiographic approach requiring the preparation of a detailed, priced, work schedule for a project. The calculation of the cumulative costs according to the project work schedule provides the cash flow profile.

Khosrowshahi (1991) described practical cash flow forecasting as 'No parameters criteria'.

> The result of this approach is the product of an extensive investigation of the specifications of each individual project. It relies, significantly, on the elemental arrangement of the cost attributes. Therefore, it consists of labouriously identifying all the cost constituents (elemental or based on arbitrary cost centres), their quantity and rate, and the order of the construction sequence of these cost components. Therefore, there exists no criterion other than the information contained within the bill of quantities and the schedule of work.

This approach, important on both epistemological and methodological grounds, has nevertheless received very little attention in the literature. It usually enjoys only a brief passing comment on the way to developing the justification of a standard curve approach. In fact, the comment very often is that the approach is the most accurate but involves a detailed analysis requiring a significant effort. The desire to avoid this effort is partly what has driven the search for standard curves over the years.

An early work (Kennedy *et al.*, 1970) developed a detailed project schedule with full costing based on the Bill of Quantities. Their model clearly showed the efficacy of developing cash flow profiles for projects from project schedules, and indeed the value of such methods for monitoring project performance during construction. However, they were very concerned about the cost of the method, which they calculated on their demonstration project cost approximately 0.96% of the total contract value; although they anticipated that better systems would reduce that cost to 0.47%. This is still a significant barrier to the method. Furthermore, they argued that a priced contractor's network was essential and 'should a realistic network not be forthcoming or not be planned in sufficient detail to be capable of providing a reasonably accurate cost analysis, the procedure cannot work' (Kennedy *et al.*, 1970). A priced contractor's network is a fully costed project schedule with all project costs allocated to the activities on the schedule.

Ashley and Teicholz (1977) developed a net cash flow model, but their method involved the preparation of an earnings curve from the construction schedule. This they did because they did not feel that a standard curve could be sufficiently flexible to handle the individual conditions of each project.

Beyond this, the research community has tended to consider the process of development of priced schedules as straightforward and not of sufficient importance to warrant detailed analysis and reporting. It is necessary to turn to operations research to find further analysis. This is only obliquely relevant here, but deals with the optimisation of Net Present Value (NPV) of cash flows for resource-constrained projects. Optimisation of NPV during construction has not been pursued in construction research, but is in fact closely related to the optimisation of net cash flow and organisational working capital. There are many papers which deal with optimisation heuristics; Padman and Smith-Daniels (1993) provide a good starting point.

THE SEARCH FOR A NEW MODEL

The idiographic–nomothetic debate flourished within the social sciences from the 1950s through to the early 1960s (Runyan, 1983). The contention arose, according to De Groot (1969), from the inability to classify the social sciences as either cultural or natural sciences. The social sciences, to which construction management must belong, have components of both cultural sciences (for example history) and natural sciences (for example physics), and have aspects which are 'individual and unique; they are own—(character)—describing: idiographic' (De Groot, 1969). De Groot claimed that 'if one seeks to conduct a scientific investigation into an individual, unique phenomenon...the regular methodology of (natural) science provides no help'. As construction projects are unique it would seem logical that their cash flows should be considered as individual and unique. The evidence in this section supports this view, and the view that an idiographic methodology has a place in the study of individual cash flows.

Individual variation between projects is caused by a multiplicity of factors, the great majority of which can neither be isolated in sample data, nor predicted in future projects. Some existing cash flow models hold that generally two factors, date and project type, are sufficient to derive an ideal construction project cash flow curve. Such convenient divisions ignore the complex interaction between such influences as economic and political climate, managerial structure and actions, union relations and personality conflicts. Many of these factors have been perceived to be important in related studies such as cost, time and quality performance of building projects (Ireland, 1983), and therefore models which ignore all these factors in cash flow research must be questioned. A number of recent studies have attempted to handle such variation.

The majority of early studies use historical data. Standard curve models, based on historical data, have been extensively used in cash flow research (for example: Kerr, 1973; Hudson and Maunick, 1974; Balkau, 1975; Bromilow and Henderson, 1977; Bromilow, 1978; Drake, 1978; Hudson, 1978; McCaffer, 1979; Tucker and Rahilly, 1982; Singh and Phua, 1984; Kenley and Wilson, 1986; Tucker, 1986, 1988; Kaka and Price, 1993). Although these approaches have gained general acceptance, they have not escaped criticism. Hardy (1970) found that there was no close similarity between the ogives for 25 projects considered, even when the projects were within one category, and Sidwell and Rumball (1982) demonstrated that the model failed in forecasting, with 30 instances where the DHSS standard model could not be shown to be a reasonable predictor in a sample of 38 projects. This implicit support for an idiographic methodology was subsequently ignored, despite the problems which some researchers found in supporting their models. Hudson observed that 'difficulties are to be expected when trying to apply a simple mathematical equation to a real life situation, particularly one as complex as the erection of a building' (Hudson and Maunick, 1974; Hudson, 1978). It is interesting to contrast the size of Hardy's sample of 25 projects, with the relatively small samples used by many of the researchers finding nomothetic, ideal curves. For example Bromilow and Henderson (1977) used four projects, while Berdicevsky (1978) and Peer (1982) used seven. Small samples should result in low confidence due to the small number of degrees of freedom, however using the individual data points in a collective analysis may provide false confidence in the model.

One group of authors (for example: Kennedy *et al.*, 1970; Peterman, 1973; Reinschmidt and Frank, 1976; Ashley and Teicholz, 1977; Berdicevsky, 1978; Peer, 1982) have modelled cash flows, prior to construction, through the use of forecast work schedules, as discussed in the previous section. Although this approach is valid for modelling predicted cash flow schedules, comparisons between the forecasts and the data for the actual project have not been published. This method of using forecast work schedules is likely to be predominantly a measure of an estimator's scheduling consistency, rather than of central trends within project groups. Historical data, on the other hand, is derived from existing projects and therefore allows direct analysis of actual, as opposed to imagined, cash flow curves.

There has been a trend over time towards an idiographic construction project cash flow model. The early models, which may be termed 'industry' models, searched for generally applicable patterns across the entire industry. When it was recognised that this was unlikely to be achieved, greater flexibility was introduced by searching for patterns within groups or categories of project (the division usually being made according to project type and/or dollar value—for example Hudson and Maunick, 1974). This still wholly nomothetic approach was modified by Berny and Howes (1982) who adapted the Hudson and Maunick (1974) category model to a form which could reflect the specific form of individual projects. Even within categories, it had been found that there were occasional projects which did not fit the forecast expenditure well (Hudson and Maunick, 1974; Hudson, 1978).

Berny and Howes (1982) designed methods for calculating the specific curve for a given project, based on their general equation. In doing so they pointed the way for future research in this field. Their model made a very important cognitive step. By proposing an equation for the general case of an individual project curve, as distinct from the curve of the general (standard) function, it moved from a nomothetic to an idiographic approach. Subsequent models have been a mix of different models for the gross cash flow (Tucker, 1986, 1988; Miskawi, 1989; Khosrowshahi, 1991; Boussabaine, Thomas and Elhag, 1999) and different, experimental approaches to modelling gross cash flow (Khosrowshahi, 1991; Boussabaine and Kaka, 1998; Boussabaine, Thomas and Elhag, 1999).

One aim of this section is to illustrate a model which, while recognising that variable influences exist, does not need to predict them in order to model cash flows. This is a model which can take on the individual shape of a project cash flow curve regardless of influences and allow the variation between projects to be examined and quantified. Having formulated the model, the idiographic approach can then be contrasted with the nomothetic approach. The search for, and testing of, such a model forms a significant portion of this section which was originally reported in Kenley and Wilson (1986).

Even as this book is being written, a debate has sprung up on this topic on the construction related CNBR email list, with some comments supporting 'standard' predictive curves (such as found in FINCASH) and other arguing that forecasting is not possible.

> I don't know if it is chaotic or because there are too many variables involved, but project cost vs time (the S curve) cannot be accurately forecasted. My conclusions are that if they cannot be accurately forecasted then let us (the contractor) prescribe what cash flow profile we want and, in order to achieve it, start negotiating with other parties (the client and subcontractors) for a possible win–win scenario (Farzad Khosrowshahi, email to CNBR, 16th January 2002).

This debate highlights the philosophical nature of the problem. Perhaps agreement will never be reached, nor indeed need it be.

The proposed model utilizes the Lorenz curve, which illustrates a relationship between two variables expressed as percentages and is common to most cash flow ogive models. It has been widely shown that the Lorenz curve expressing the relationship between percentage of project time completed (as the independent variable on the abscissa) and cumulative value at any time as a percentage of final value (the dependent variable on the ordinate), takes on a sigmoid form. Figure 3.7 illustrates an example project data set for cash flow from a client to a contractor with a suggested fitted curve.

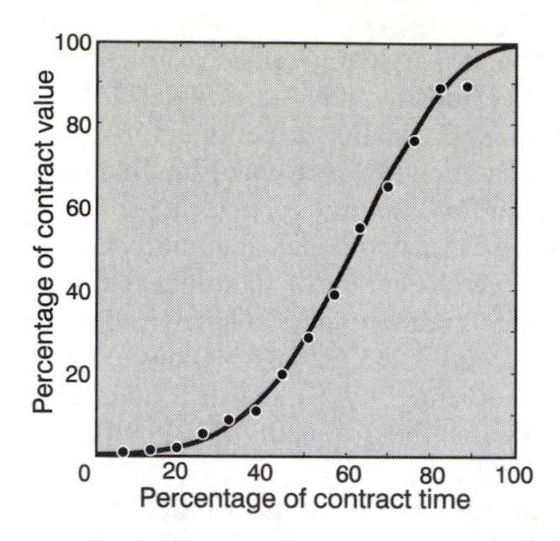

Figure 3.7 Example of project data set with fitted curve (Kenley and Wilson, 1986)

A second requirement of this model is to allow for the subsequent modification of the model to a net cash flow model. Nazem (1968) suggested that cash balance curves could be obtained indirectly from the inward and outward cash flow curves. With two such curves generating each net cash flow curve, the problems inherent in a nomothetic methodology are accentuated. The net cash flow model will be presented in Chapter 6.

The term 'cash flow' has been defined as describing the flow of cash or commitment from the client to the contractor. The model is equally applicable to flows from the contractor to his various subcontractors and suppliers. An important clarifying distinction must be made between cash flow and net cash flow (cash balance), the latter being the residual after both income and outgoings are assessed. The model outlined here is equally capable of handling the value of work in place, the value of progress payments claims, or the value of progress payments actually paid to the contractor, despite the differences in time and monetary value which each of these may have for a given project.

The model fits the points of payment. On a stepwise graph of gross cash flow, the model fits the upper points as illustrated by line A in Figure 3.8. The use of the model to fit other situations is discussed in Chapter 6.

The principles of regression analysis and independent samples

A significant rationale for the introduction of an idiographic methodology is based upon a consideration of the principles of regression analysis. Researchers modelling cash flows under a nomothetic methodology have utilized polynomial regression as their curve-fitting technique. Given sufficient constants and terms it seems possible to fit a polynomial to any required sigmoid form. However, the nomothetic regression operates upon grouped data, a procedure which involves overlaying the data from many projects. Given the nature of the data involved and the principles of regression analysis, the validity of this procedure is questionable.

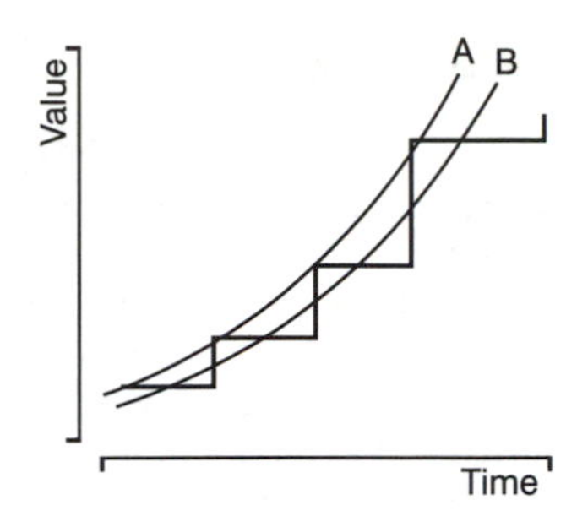

Figure 3.8 Stepped inward cash flow with top point trend line

The principles of regression analysis assume that dispersion of points about a central tendency is normally distributed and caused by random error. Spirer (1975, p. 329) states the assumptions applicable to a relationship in a bivariate population to be:

1. The conditional mean, $\mu_{Y,X}$ is related to the independent variable *Y*.
2. The conditional standard deviation, $\sigma_{Y,X}$ is the same for all values of the independent variable X.
3. The conditional distribution of *Y* for any *X* is modelled by the normal probability distribution.

The regression model therefore presupposes no relationship between *Y* (dependent) values except for their dependency upon *X* (independent) values.

In reality, discrepancy from the average trend line is caused by systematic, idiosyncratic variation in the profile for each project, rather than random error. The *Y* value for each point is specifically linked to the preceding or succeeding points for that project, whilst maintaining no relationship whatsoever to any other data points from other projects in the analysis. Thus each project has an individual line of central tendency, with an associated error scatter about the line. A group regression is only applicable if it can be shown that there is no significant difference between the individual conditional means $\mu_{Y,X}$, and therefore no significant difference between the lines of best fit. The evidence here shows that this is fundamentally incorrect.

The only situation in which regression may be appropriate is for the analysis of the trend line of a single project, where the hypothesis that all deviation from the central mean is caused by random error may reasonably be accepted. It follows then, that if regression analysis is to be used it cannot correctly analyse more than one project at a time. It must be said that, from the literature, the research community remains unconvinced.

> I shall say nothing about philosophy, but that, seeing that it has been cultivated for many centuries by the best minds that have ever lived, and that nevertheless no single thing is to be found in it which is not subject of dispute, and in consequence which is not dubious, I had not enough presumption to hope to fare better there than other men had done. And also, considering how much conflicting opinions there may be regarding the self same matter, all supported by learned people, while there can never be more than one which is true, I esteemed as well-nigh

false all that only went so far as being probable (Descartes, cited in Rescher, 1985).

There are other problems stemming from the use of polynomial regression. First a large number of constants are generally required. Peer (1982) used a biquadratic equation with five constants; Berdicevsky (1978) used a cubic with four; so also did Bromilow and Henderson (1977), whose inverted polynomial expressed time as a function of value. Hudson (Hudson and Maunick, 1974; Hudson, 1978) used a cubic polynomial, rearranged to contain only two variable constants. In addition a full polynomial regression is necessary for each fit, and the lines of best fit derived do not always intersect the origin (0,0) and end (100,100) (which is an advantage for a cash flow model intended for modification to a net cash flow model).

There is an alternative to the use of polynomial regression analysis for curve fitting. This is linear transformation, which involves altering the values of either the independent, the dependent, or both, data variables by a mathematical function in order to render a linear relationship between the transformed data variables. The methods for choosing transformations require specific knowledge about the relationship between variables, or use diagnostics to suggest possible transformations (Weisberg, 1980).

THE KENLEY AND WILSON LOGIT MODEL

The cash flow ogive is a sigmoid, and similar curves have been found in connection with growth patterns in economics and biological assay. Investigations in these areas have found that specific transformations of sigmoid curves can produce linear functions. The parameters of the sigmoid function are provided by the parameters of the linear equation, which in turn are found through linear regression.

The equation of the curve of best fit for data which approximates one of the sigmoid growth functions may be found by a linear regression of suitably transformed data, and then substitution of the linear parameters into the sigmoid function. The result may then be tested for goodness of fit against the data.

The selection of an appropriate sigmoid function in bioassay was examined by Ashton (1972), who outlined four of the best known sigmoids and their transformations; the integrated normal curve, the logistic curve, the sine curve and Urban's curve. Ashton found the above curves to be very similar in shape with almost all variation seen only at the extremes. He therefore concluded that the selection of an appropriate sigmoid was more a matter of application than anything else. If it could be shown that the ontological form of the cash flow, described by Hudson and Maunick (1974) as the underlying relationship between expenditure and time, approximates the form of the sigmoids listed above, then it would follow that any of these sigmoids could be used. This has been found to be so, although it is necessary to use a double transformation in cash flow analysis.

The Logit (given in Equation 3.1) is the simplest of the Ashton sigmoid transformations and most easily allows the change to double transformation. The linear equation is found by a logit transformation of both the independent and dependent variables:

$$Logit = \ln\left(\frac{z}{1-z}\right) \tag{3.1}$$

where z is the variable to be transformed, and *Logit* is the transformation. The logistic equation for cash flows can be expressed using value v as the dependent variable and time t as the independent variable:

$$\ln\left(\frac{v}{1-v}\right) = \alpha + \beta X \text{ where } X = \ln\left(\frac{t}{1-t}\right)$$

therefore

$$\ln\left(\frac{v}{1-v}\right) = \alpha + \beta\left(\frac{t}{1-t}\right) \tag{3.2}$$

Equation 3.2 then forms the equation of the sigmoid curve that describes the flow of cash on a specific building project. It may also be expressed in terms of v as follows:

$$v = \frac{e^{\alpha}\left(\frac{t}{1-t}\right)^{\beta}}{1+e^{\alpha}\left(\frac{t}{1-t}\right)^{\beta}} \tag{3.3}$$

$$\text{or } v = \frac{F}{1+F} \text{ where } F = e^{\alpha}\left(\frac{t}{1-t}\right)^{\beta} \tag{3.4}$$

The Logit cash flow model given in Equation 3.4 uses scales from 0.0 to 1.0, where the ratio (on the abscissa or ordinate) 1.0 is equivalent to 100%. As percentage scales are to be used, in accordance with convention, the equations should be expressed as follows:

$$\text{If } \ln\left(\frac{v}{100-v}\right) = \alpha + \beta\left(\frac{t}{100-t}\right)$$

$$\text{then } v = \frac{100e^{\alpha}\left(\frac{t}{100-t}\right)^{\beta}}{1+e^{\alpha}\left(\frac{t}{100-t}\right)^{\beta}} \tag{3.5}$$

$$\text{or } v = \frac{100F}{1+F} \text{ where } F = e^{\alpha}\left(\frac{t}{100-t}\right)^{\beta} \tag{3.6}$$

The practical application of the Logit transformation cash flow model implies that construction project cash flow curves approximate the sigmoid form yielded by Equation 3.6. This being so, a transformation of the data should approximate a line described by Equation 3.7, and with parameters α and β:

$$Y=\alpha+\beta X \text{ where } Y=\ln\left(\frac{v}{100-v}\right) \text{and } X=\ln\left(\frac{t}{100-t}\right) \tag{3.7}$$

where V=100% or 1.0 and T=100% or 1.0 for percentage or ratio data respectively. The Logit transformation model is represented by Equations 3.4, 3.6 or 3.7.

In Figure 3.9 the data for a sample project is illustrated first in the Lorenz format, and secondly as transformed. From this it can be seen that the transformed data can indeed be approximated by a straight line.

In order to transform the data, X and Y must be calculated for each value of t and v respectively. Deriving the constants α and β is then simply a matter of linear regression of the transformed data, where:

$$\beta=\frac{\sum\left[(X-\overline{X})(Y-\overline{Y})\right]}{\sum(X-\overline{X})^2} \tag{3.8}$$

$$\text{and } \alpha=\overline{Y}-\beta\overline{X} \tag{3.9}$$

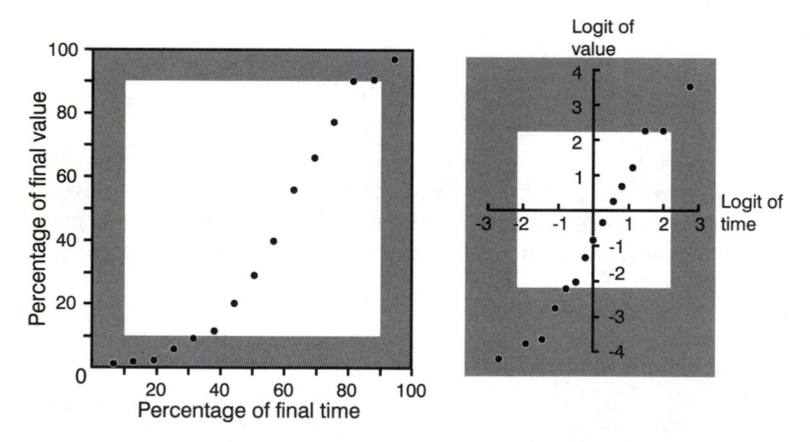

Figure 3.9 Linear transformation for a sample project (Kenley and Wilson, 1986)

Aims for the Logit model

In summary, the following rationale for an idiographic methodology for construction project cash flows have been presented:

1. Consideration of the idiographic–nomothetic debate in the literature led to the conclusion that the natural science methodology was inappropriate for unique phenomena such as construction projects.
2. A multiplicity of factors and influences affect project cash flows, many of which are unquantifiable and have differential impact.
3. Regression analysis for grouped data, associated with a nomothetic methodology, is of questionable validity for construction projects.

It is therefore contended that an idiographic methodology is more appropriate to the study of construction project cash flows than is a nomothetic methodology, and a nomothetic methodology can only be supported if a significant similarity can be shown to exist within groups. The experimental hypothesis is that there is substantial variation between projects.

The principal aim is to develop an idiographic construction project cash flow model. Secondary aims are:

1. To determine an optimum exclusion range for data points;
2. To examine the model for goodness of fit and to identify projects where the model has failed to adequately fit the data (referred to as outliers);
3. To use the model to support the hypothesis above, by demonstrating variation between projects;
4. To support the contention above by contrasting the two methodologies quantitatively; and
5. To allow for the subsequent development of a net cash flow model.

Measures of variance

One area of importance which is touched on by most authors in their need to measure the performance of their model, is the measure of the goodness-of-fit.

In order to draw conclusions within models and comparisons between models, it is necessary to develop a measure of the goodness of fit. The measure most often chosen, put forward as a risk index by Jepson (1969) and given the acronym 'SDY' by Berny and Howes (1982), is the standard deviation about the estimate of Y. *SDY* adopts the common measure of dispersion.

$$\sigma = \sqrt{\text{var}} = \sqrt{\frac{\sum(X-\overline{X})^2}{N}} \tag{3.10}$$

The fitted model is then declared to be the measure of central tendency or conditional mean of the Y values of the data. Therefore $\overline{X}$ is replaced by the estimated (or fitted) value of $Y(Y_E)$ for a given X coordinate. The variance being measured is of Y about Y_E.

The rationale underlying this step is that for each point X there is a true mean value of Y and that any deviance can be explained by random, normally distributed error. A systematic error would imply a limitation of the fitted model. The measure of dispersion is therefore:

$$SDY = \sqrt{\frac{\sum (Y - Y_E)^2}{N}} \tag{3.11}$$

This measure is suitable for inter-model comparison, where the model with the lowest *SDY* value demonstrates the best fit and is therefore the most desirable model. Boussabaine and Kaka (1998) adopted a root–mean–square error (RMS) measurement which is identical to the SDY.

There are two problems with the *SDY* measure, the first is that it is dependent on the scale of the observed variable, and the second is its failure to accommodate degrees of freedom.

The SDY model works as long as data is normalised in or dependent of the scale, for example as percentage data. If the data cannot be expressed in percentage terms, then the formula may be adjusted by dividing SDY by the maximum value or a base figure. This is the same effect as normalising in the formula rather than all the data. The complete formula is:

$$SDY = \frac{\sqrt{\frac{\sum (Y - Y_E)^2}{N}}}{Y_{\max}} \tag{3.12}$$

As this gets the same results, the short form will be used from here on.

Skitmore used MSQ (Mean Square Error) as his measure. This differs from *SDY* in that it provides for the degrees of freedom in the model as well as being the square of SDY.

$$MSQ_{(a,b)} = \frac{\sum_{i=1}^{n} (v_i - u_i)^2}{n - 2} \tag{3.13}$$

This corrects a minor error in the work of Kenley and Wilson (1986) and Kaka and Price (1993). However, Skitmore has made a similar error with the use of two degrees of freedom. The concept of 'degrees of freedom', is described by Wonnacott and Wonnacott (1972:167) in the following intuitive way:

> Originally there are n degrees of freedom in a sample of n observations. But one degree of freedom is used up calculating $\overline{X}$, leaving only (n – 1) degrees of freedom for the residuals $(X - \overline{X})$ to calculate s^2 [variance].

This argument is continued for simple regression (page 273):

> ...two estimators $\hat{\alpha}$ and $\hat{\beta}$ are required; thus two degrees of freedom are lost for s^2 [variance]. hence (n – 2) is the divisor in s^2.

This argument means for cash flow modelling that calculation of internal validity through a linear regression procedure requires two points and thus uses two degrees of freedom. The logical extension of this approach indicates that polynomial regression provides fewer degrees of freedom. Thus, for example, the Peer (1982) Equation (3.20) has four terms and would use four degrees of freedom. Skitmore used only two degrees of freedom for all equations with his MSQ.

It can further be argued that a comparison between models requires no points and therefore has *n* degrees of freedom and accordingly the correct divisor is *n*. Thus the correct calculation of variance in testing the validity of a model is summarised in Table 3.3.

Table 3.3 Forms of the equation for variance with degrees of freedom

Context	Degrees of freedom	Formula	Equation #
Between models	n	$SDY=\sqrt{\frac{\sum(Y-Y_E)^2}{n}}$	(3.15)
Within models—Linear regression	$n-2$	$SDY=\sqrt{\frac{\sum(Y-Y_E)^2}{n-2}}$	(3.16)
Within models—4th degree polynomial regression	$n-4$	$SDY=\sqrt{\frac{\sum(Y-Y_E)^2}{n-4}}$	(3.17)
Within models—$(n')^{th}$ degree polynomial regression	$n-n'$	$SDY=\sqrt{\frac{\sum(Y-Y_E)^2}{n-n'}}$	(3.18)

Tucker (1986) uses the Chi Square goodness-of-fit statistic. 'Chi Square (χ^2) is a very popular form of hypothesis testing, and one that is subject to substantial abuse' (Wonnacott and Wonnacott, 1972: 423). The formula:

$$\chi^2=\sum_{i=1}^{k}\frac{(O_i-E_i)^2}{E_i} \tag{3.14}$$

is intended for frequency distributions and not for instances where numeric data is available. It is therefore not a suitable measure for measuring the accuracy of cash flow models.

Method

Data

The model was tested on data from two separate, and different sources.

The first sample (S1) comprised data for 32 medium- to-large scale commercial and industrial projects, provided by a Sydney-based construction group. The requirements of anonymity prevented the provision of a detailed breakdown into subcategories, as the only data provided were the dollar amounts of monthly claims from the client and the dates of approval. The projects were all constructed in the mid-to-late 1970s, in and around the Sydney and New South Wales area.

The second sample (S2) comprised data for projects from throughout Australia, provided by the Melbourne office of a quantity surveying firm. The sample included many projects, of which 40 had sufficient data for analysis. (S2) differed from (S1) in that it covered a wide range of contractors, using different contractual arrangements. The data included the type of construction, which allowed for a more detailed breakdown into subcategories. The projects were constructed in the period from the late-1960s to the mid-1970s.

Exclusion of data at the extremes

The nature of the logit transformation is such that as the data approaches either 0% or 100%, then the logit will approach positive infinity or negative infinity respectively. One of the limitations of linear regression is that any extreme values will dominate the analysis, to the extent that α and β values might be unduly influenced by a small number of extreme values rather than by the bulk of the data.

Within cash flow analysis, the extreme data points are arguably the least significant, whereas they are the most dominant in the regression analysis. A simple method for countering the problem is to exclude the data points outside an acceptable range from the analysis. An appropriate cutoff percentage for end point data (to be referred to as the exclusion range) will be found by trial and error, using the *SDY* as an indicator.

Design and procedure

The conventions used when collecting and analysing the data in the analysis are set out below so that comparisons may be made, and the tests repeated. Time was measured in calendar days. It was assumed that the Lorenz value ogive for working days would approximate that for calendar days. This approach ignores the effect of the substantial holiday commonly taken in January by the Australian construction industry, which may result in discontinuities in the ogives for specific projects. As project data for days worked was not available, this crude assumption was considered necessary and the results indicate it was adequate.

The base figure for 100% value was taken to be the final amount certified; the equivalent figure for time was taken to be the time at which 100% value was certified This point was considered preferable to the date of practical completion sometimes proposed (for example by Balkau, 1975; Bromilow, 1978; Bromilow and Henderson, 1977; Tucker and Rahilly, 1982) because the latter is irrelevant to a flow of cash and would be meaningless in a net cash flow model. The origin was taken to be the commencement of work on site, a convenient and easily identified point in time. All projects have an origin and an end, and thus a model which incorporates these points would be advantageous.

The procedure for testing the model involved the calculation of α and β values. The SDY measure was used as an indicator of comparative accuracy. When this research was first published (Kenley and Wilson, 1986), the SDY value for the model was calculated using *n* degrees of freedom. This was most likely an error and the degrees of freedom for the measure of internal consistency should have been *(n – 2)*. However, to avoid confusion, the *SDY* values in this section have not been changed from the original analysis.

A systematic trial and error process was used to locate the optimum exclusion range. The analysis was run 31 times for each project in each sample (S1 and S2), excluding from the 0% to the 30% ranges in steps of 1%. The lowest mean sample *SDY* value indicated the optimum exclusion range. The mean sample *SDY* could then be used to compare the two sample groups using a non-parametric test—the W test (Wonnacott and Wonnacott, 1972).

The distribution of the *SDY* values for the projects, with extreme data excluded, formed the basis for a further non-parametric test (the Boxplot test available on MINITAB—see Hoaglin, Mosteller and Tukey, 1983; Velleman and Hoaglin, 1981) designed to identify outliers in the population, which are projects for which the model does not fit the data with a statistically acceptable *SDY.*

A random sample of profiles was selected from each of S1 and S2 to demonstrate the diversity and extent of variation attained from the data, and thus the variation between projects.

Finally, the results of a traditional nomothetic standard curve analysis were compared with the results of the idiographic analysis. In this experiment an average curve was calculated for each of S1 and S2. Then the *SDY* for each project, based on the average curve as the line of central tendency, was calculated and contrasted with the individual *SDYs* already achieved.

In summary the following experiments are reported in the following order:

1. Testing for optimum cutoff point using a trial and error basis and mean sample *SDY* as an indicator (with *n* degrees of freedom);
2. Examining the samples for significant differences which would indicate they were from separate based populations;
3. Using the model derived from (i) to identify and remove projects which may be considered outliers according to a non-parametric test;
4. Examining the influence of inflation on project *SDY* for the model through the use of a Building Cost Index which tests the prediction that the model is equally good for differing data conventions;
5. Assembling a random selection of profiles from S1 and S2 and inspecting for variation in the fitted models of individual projects; and
6. Comparison of deviation from a calculated average curve with deviation from a project specific curve using the *SDY* measure, to demonstrate the failure of a group average model.

Results and discussion

The results of experiment (1) proved to be very significant. Figure 3.10 illustrates the mean sample *SDY* values for the trial and error analysis, and the graph shows that a lower mean sample *SDY* is achieved if the upper and lower 13% (Sl) or 11% (S2) extremes of data are excluded from the analysis. It was decided to choose 10% as the optimum exclusion range, because the curves reach a minimum from about 10% to 16% and it is desirable to select the percentage which utilizes as much of the data as possible, without affecting results detrimentally. Table 3.4 compares the mean *SDY* and standard deviation values for the model for all data, and for the 10% exclusion range. These figures show clearly that not only are the mean sample *SDY* values lower, but the standard deviation of the mean sample *SDY* also decreases markedly at the 10% exclusion range. At the 10% exclusion range the model is not only a better fit but has a smaller range of *SDY* values. Fewer projects are thus likely to be excluded as outliers by the *SDY* measure.

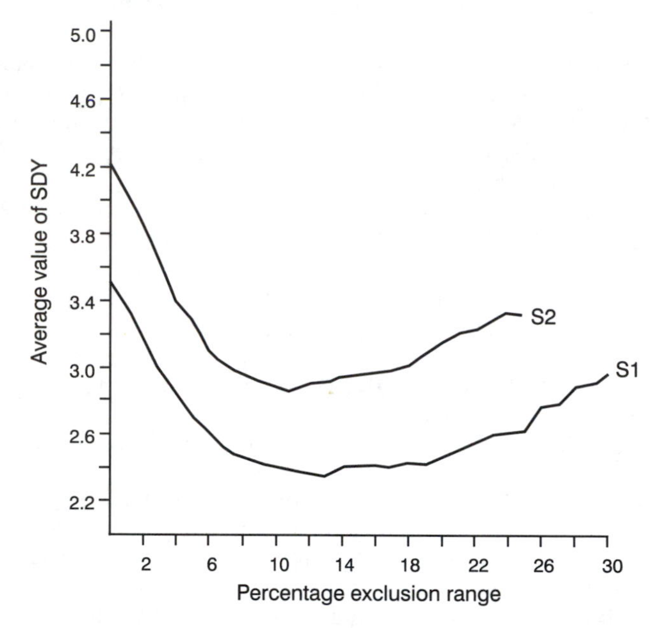

Figure 3.10 Mean sample SDY against exclusion range
(adapted from Kenley and Wilson, 1986)

The lower mean sample SDY and standard deviation for S1 as compared with S2 prompts the question, is the model achieving significantly different results for each sample? In other words, are the samples statistically from the same or different populations. This was tested in experiment (2) by the use of a non-parametric W-test (Wonnacott and Wonnacott, 1972), chosen for its ability to cope with suspected outliers. The test examined whether the two underlying populations were centred differently, or were one and the same. The results of the W-test indicated that with a 95% confidence the samples were from the same population. This result was supported by a t-test. Thus there is no support for the treatment of the results from the model as coming from independent samples.

Table 3.4 Mean sample SDY for samples S1 and S2 (adapted from Kenley and Wilson, 1986)

Sample	Exclusion range (%)	Mean sample SDY	σ
S1	0	3.52	1.61
	10	2.40	0.81
S2	0	4.23	2.35
	10	2.88	1.30

Given that the samples could be treated as one, the samples were examined in experiment (3) for the existence of outliers. Another nonparametric test, the Boxplot (MINITAB), found no outliers in Sl, but two possible outliers in S2. It can therefore be concluded that of the 72 projects tested, only two failed to be fitted adequately. Figure 3.11 illustrates the Boxplot output for S1 and S2 combined; the dots identify outliers. Effectively, an *SDY* of 6% can be deemed the determinant *SDY* value for outliers, and thus these are projects for which the model appears unreliable.

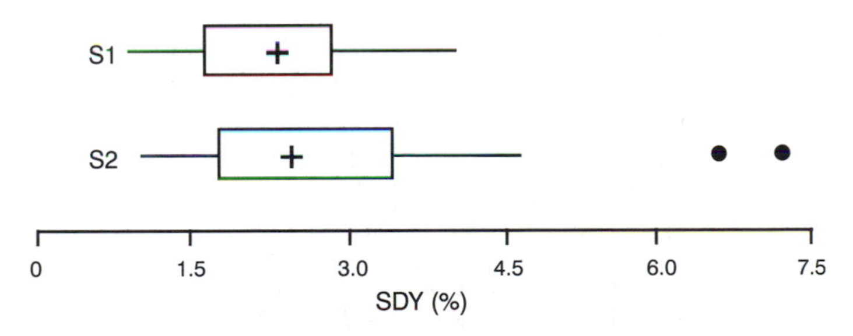

Figure 3.11 Boxplot of project SDY values for samples S1 and S2 (Kenley and Wilson, 1986)

The visual inspection required for experiment (4) supports the experimental hypothesis. Figure 3.12 illustrates a random sample of actual project curves produced by the model in this analysis, and demonstrates the wide range of profiles possible.

The 'ideal sigmoid' is an illusion; cash flows may be concave or convex as well as the more common 'S' shape. The projects vary in slope and lag, and some demonstrate a prolonged start or conclusion; one curve was an inverted sigmoid; whilst others do approximate to the ideal curve reasonably well. The curve envelope is very wide and includes a fascinating range of potential cash flow curves.

As the sample demonstrated a wide range of profiles, the α and β constants for each project have been included in Table 3.5 (page 58), so that the graphs may be reproduced by an interested researcher. These findings support inferences which may be drawn from the work of Singh and Phua (1984) whose graphs illustrate a high degree of inter-project variability.

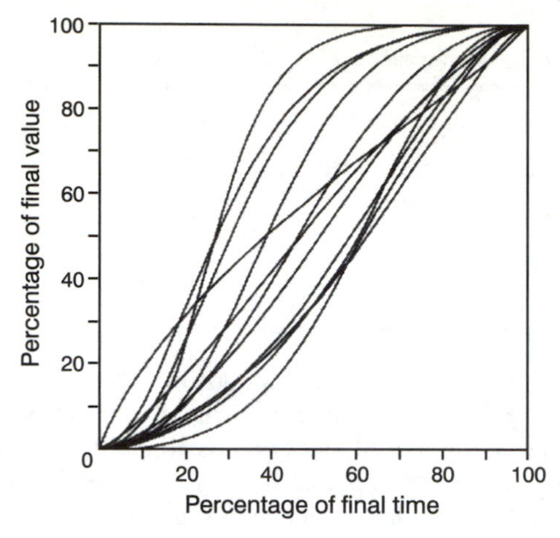

Figure 3.12 Random sample of project curves (Kenley and Wilson, 1986)

A powerful argument in support of the idiographic methodology is provided by experiment (5). Figures 3.13 and 3.14 illustrate the scatter plots for all projects in S1 and S2 (outliers were excluded from this analysis).

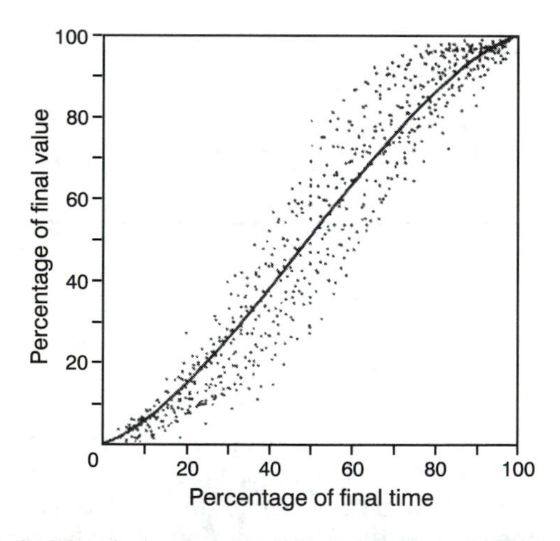

Figure 3.13 Standard curve analysis for S1 (Kenley and Wilson, 1986)

The lines shown are the lines of best fit for each sample. A casual inspection might find the results to be in support of a nomothetic approach as the fitted lines appear as a good average of all the points (although skewed at the extremes). This conclusion would be misleading however as it does not consider originating data. It should also be noted that the best fit lines for this experiment excluded data in the exclusion range, a technique designed to optimize fit for individual projects. The resultant fit is poor in this range explaining the apparent skew in the figures.

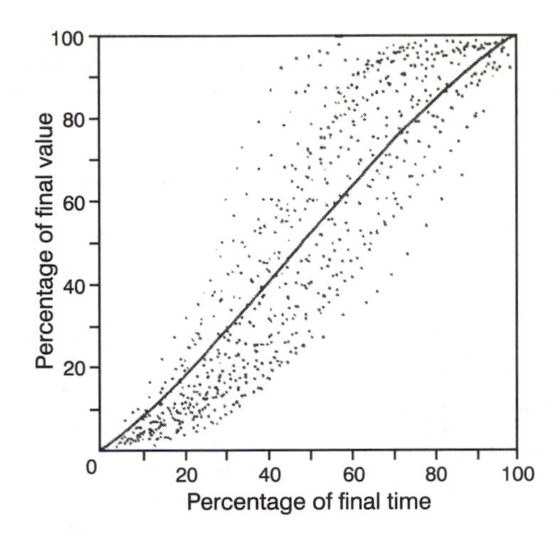

Figure 3.14 Standard curve analysis for S2 (Kenley and Wilson, 1986)

It can be seen from Figure 3.12 that the lines which connect the points, while passing through the overall envelope of possible points, can do so in significantly different ways, and thus the average curve is misleading.

The contrast between the average curve and the individual curve is demonstrated by the results shown in Table 3.6. This shows the *SDY* values for each project, first, based on the average curve as the central tendency, and secondly, based on the individual curve as the central tendency. In all but one case the *SDY* value for the general model is higher than the *SDY* value for the specific model. In 80% of projects (69% for S1 and 89% for S2) the fit for the general model exceeds the *SDY* value set as the determinant for outliers for the specific model. In other words the fit of the general model is only unacceptable for 20% of projects. These results support the contention made, in the discussion of the principles of regression analysis, that project data deviates from a central mean by a combination of both systematic (resulting from the individual ontology of the project) and random error.

General discussion

An idiographic construction project cash flow model has been developed and successfully tested within this section. The model has achieved excellent results over large samples, and only two of 72 projects were rejected by the model. The variation from the model was restricted to an *SDY* ranging from 1.0% to 4.6% (excluding outliers) for individual projects, with a median of 2.5% approximately (including outliers), which suggests that there is little systematic error left in the model.

In contrast the results of experiment (5) suggest that random error is insufficient to explain variation in a nomothetic model, which must therefore contain substantial systematic error. The later section 'Detailed review of alternative cash flow models' (see page 61) will compare the goodness-of-fit for the model with other models.

Table 3.5 α and β for S1 and S2 with associated SDY (Kenley and Wilson, 1986)

S1			S2		
α	β	*SDY*	α	β	*SDY*
0.94	2.13	1.65	–0.82	1.83	1.81
–0.20	1.98	2.89	–0.57	1.27	1.60
0.07	1.24	1.57	–0.69	1.51	3.23
0.37	1.46	3.65	0.36	2.30	3.42
–0.15	1.43	2.72	0.23	3.86	4.09
0.13	1.38	3.71	–0.98	1.45	2.72
0.38	0.84	3.69	1.10	2.12	2.26
–0.31	1.34	1.42	2.83	2.86	2.63
0.36	1.26	1.03	0.77	1.88	1.88
–0.23	1.25	2.77	–0.05	1.67	1.55
0.42	1.72	1.92	0.69	2.09	4.10
0.25	1.30	2.41	0.65	1.80	3.38
0.23	1.71	3.16	1.58	1.94	4.66
–0.16	1.65	1.85	0.67	1.85	*7.33
–0.25	1.18	2.37	–0.38	1.52	1.91
0.52	1.78	1.28	0.44	1.81	1.97
0.68	2.35	3.15	–0.82	1.13	2.18
–0.54	1.50	2.47	0.86	2.15	2.27
–0.69	1.63	4.11	1.76	1.80	3.52
1.00	1.75	2.26	–0.53	1.34	2.56
–0.19	1.46	2.86	0.65	1.34	2.64
0.50	1.35	2.00	–1.05	0.96	3.69
–0.76	1.21	2.66	1.81	2.40	4.29
–0.68	1.30	1.86	0.66	1.53	2.68
0.00	1.26	1.32	–0.10	1.44	2.26
0.29	1.44	1.74	–0.29	1.58	2.24
0.59	1.74	1.90	1.27	1.78	1.64
–0.51	1.39	1.85	–0.27	1.43	1.42
–0.24	1.24	2.42	–0.83	1.51	1.87
–0.90	2.01	1.72	0.06	2.40	1.92
0.76	1.31	3.51	0.14	1.22	2.34
0.73	1.71	3.05	0.69	1.77	3.50
			0.12	1.16	3.55
			1.41	1.91	*6.80
			–0.19	1.06	3.76
			0.81	1.72	2.70
			–0.66	1.31	1.22
			–0.04	1.81	2.19
			0.83	2.44	2.73
			0.71	1.46	2.70

*Possible outlier.

Table 3.6 Comparison of SDY
for individual and average models for each sample

S1		S2	
Average	Individual	Average	Individual
13.60	1.65	14.38	1.81
7.99	2.89	11.24	1.60
1.97	1.57	12.42	3.23
6.38	3.65	12.77	3.42
3.82	2.72	17.40	4.09
5.29	3.71	15.64	2.72
11.55	3.69	17.30	2.26
5.46	1.42	29.82	2.63
6.41	1.03	12.71	1.88
4.10	2.77	6.27	1.55
6.83	1.92	13.65	4.10
4.55	2.41	12.19	3.38
6.41	3.16	21.90	4.66
5.08	1.85	8.14	1.91
5.17	2.37	8.71	1.97
8.43	1.28	15.35	2.18
12.01	3.15	14.58	2.27
9.33	2.47	24.45	3.52
12.23	4.11	10.16	2.56
14.36	2.26	11.04	2.64
4.44	2.86	20.93	3.69
9.03	2.00	24.69	4.29
13.19	2.66	10.34	2.68
11.12	1.86	4.80	2.26
1.43	1.32	7.59	2.24
4.75	1.74	18.06	1.64
9.44	1.90	6.77	1.42
8.00	1.85	15.41	1.87
4.62	2.42	10.90	1.92
12.94	1.72	2.91	2.34
13.12	3.51	12.21	3.50
11.86	3.05	3.48	3.55
		4.43	3.76
		13.23	2.70
		12.10	1.22
		7.87	2.19
		14.66	2.73
		12.14	2.70

Several rationales for the introduction of the idiographic methodology have been presented, the first of which is philosophically based. The nomothetic approach assumes there are consistent similarities between projects and the proponents of this approach, then produce what are viewed as non-transient industry averages for groups of projects. This rationale ignores idiosyncratic differences between projects, discounting their significance by treating such variation as random (hence implying unimportant) error. If there exist for groups of projects discoverable, consistent, non-transient industry averages, the error component of which is truly random, then the prevailing usage of the nomothetic models, for predictive purposes, is justified. Conversely, if the above assumptions are violated, nomothetic prediction is invalid and probably meaningless.

The conditions required (for predictive purposes) are not fulfilled. Variation between projects is a product of their individuality rather than a random error about an established ideal. Therefore there is no such thing as an ideal curve for groups. Furthermore, the process of updating ideal curves, by the recalculation of parameters at intervals, is simply the selection of a new sample—and not an allowance for time.

The contention that an idiographic methodology is more appropriate than a nomothetic methodology for the study of construction project cash flows has therefore been supported. The projects examined have yielded individual profiles, which support Hardy's (1970) contention that no close similarities exist between projects. It is the authors' belief that group models are functionally and conceptually in error.

The above conclusion may have repercussions within the industry. Nomothetic models have achieved widespread support, especially where forecasting is concerned. However, consistent non-transient industry averages, with only a random error component of deviation, have not been found in the present work. Therefore standard or average curves do not reflect the individual projects, and cannot be used for forecasting.

The industry has expressed a need to predict construction project cash flows, and will continue to do so, despite any arguments that present methods are invalid. The desire of the industry to forecast cash flows is entirely understandable. Fortunately the construction industry already deals with such uncertainty in its everyday operations. Elaborate methods have been established to deal with such 'fuzzy' areas as time and cost estimating, and bidding. The role of the estimator is essentially that of predicting the unpredictable, and therefore encapsulates cash flow estimating.

To suggest that estimating differs in some way from forecasting and prediction may appear pedantic, but is fundamental to the debate at hand. The idiographic model provides a tool for the estimator, as indeed does the nomothetic model, both of which can be combined with personal judgement to arrive at an estimate of future cash flows. There can be no objection to an estimator predicting cash flows based on the evidence available, providing that the limitations of this subjective decision process are recognized (for example that extrapolation from a nomothetic model is uncertain). There will always be a place for personal judgement within estimating in the building industry. In the next section, some examples of models designed specifically to accept and manage this fuzzy nature will be introduced.

The Logit cash flow model is extremely flexible mathematically. It allows for the development of a net cash flow model. It is equally functional under nomothetic and idiographic methodologies, and therefore can be used by the estimator in whatever form required.

In conclusion, the Logit transformation construction project cash flow model has been found suitable and accurate in idiographic post hoc analysis of project cash flows. It is its ability to model the individual form of projects that also makes it suitable for comparitive analysis of other cash flow models.

DETAILED REVIEW OF ALTERNATIVE CASH FLOW MODELS

This section revisits the literature on cash flow modelling, to discuss in greater detail the previous models, to present the cash flow profiles derived and to compare to the Logit cash flow model. In this way the mathematics of each model can be provided, the resultant curve shown and an equivalent logit model developed—with its accuracy determined by the SDY indicator, using Equation 3.15 for comparison between models with n degrees of freedom.

The Bromilow model

The Bromilow model was developed at the Division of Building Research, Commonwealth Scientific and Industrial Research Organisation (CSIRO) in the 1960s. Although the model has become an important tool in the Australian industry, Bromilow has not published the method or results of his analysis, but has used the conclusions to illustrate his later work. The purpose of this work was to achieve the forward planning of large programs of building works in the normal operation of Government works departments and similar institutions. A computer model was to be generated incorporating much related research being, then and later, undertaken by CSIRO. The model of an industry-based standard curve was subsequently accepted as established and progressively updated (Balkau, 1975; Bromilow, 1978; Bromilow and Davies, 1978; Tucker and Rahilly, 1982, 1985). Work based on that of Bromilow and his colleagues has become widely known as the Bromilow model.

Part of the Bromilow model included an estimate of cumulative payments over time on individual projects. Prior to 1969 it was often held, despite Jepson's comment to the contrary (Jepson, 1969: 37), that the relationship could be approximated as linear, implying that progress payments were of equal value and at equal intervals. Bromilow examined four medium sized industrial and commercial projects, and found the cumulative relationship to be *S* shaped rather than linear, which was consistent with the work of other researchers in this and related fields (for example Pilcher, 1966). This finding was an important contribution and is generally accepted by later researchers as having been proven.

Bromilow used polynomial regression (least squares) to find an equation of best fit for the curve. An inverted cubic function $X = f(Y)$ was found to have the smallest residual variance, and as no systematic differences were found between the four projects only one curve was generated.

The equation derived is of the form:

$$T = C_0 + C_1P + C_2P^2 + C_3P^3 + C_4P^4 \tag{3.19}$$

Where T is the percentage of construction time since start to practical completion, P is the target cumulative payment expressed as a percentage of planned cost, and C_s are constants.

An early equation had constants (Balkau, 1975): C_0 =10.667, C_1 =2.09132, C_2 =−0.034436, and C_3 =0.00024155 which gave a 10.7% lapse of time before any payments were made. This shows as 12.72% time for 1% value on Figure 3.15. It also provided 100% value to be reached at 117% of planned duration. This reflected the previous work of Bromilow which had identified that project time could be expected to over-run by, apparently, 117%.

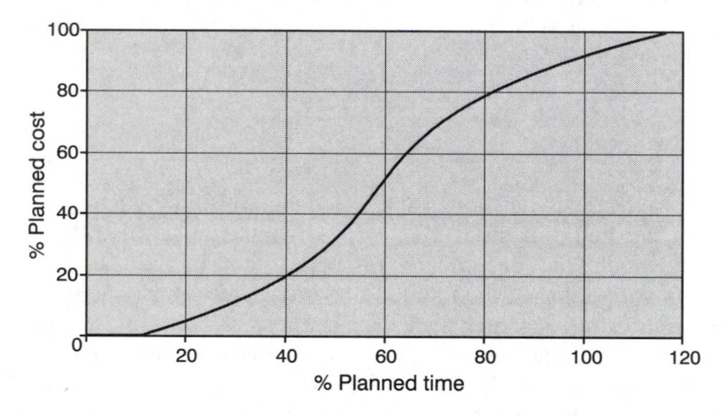

Figure 3.15 Bromilow's S curve using Balkau (1975) constants

There are properties which are inherent in the Bromilow model. These are apparent when the resultant curve is converted to the Logit model. The Bromilow model does not start at (0,0) or finish at (100,100). These properties reduce the capacity of the Logit model to provide a good fit, however the fitted model shown in Figure 3.16 has an SDY of only 2.43% and, being a Logit model, has the enormous advantage of expressing cost as a function of time.

It can be seen that the Logit equivalent, with α= 0.132 and β= 1.822, is a reasonable approximation to the Bromilow equation for forecasting purposes. Figure 3.15 replicates a chart very frequently used by estimators for graphical forecasting of cash flows (although estimators often use a version with a maximum 100% time scale). The Logit equation parameters allow this practice to continue, but with the advantage of being able to calculate cost given time, which is necessary for forecasting progress claims.

Whilst a starting value of t=10.667 may have been realistic on small projects, it was not so on large projects. So the constants were amended and the following constants were provided, with a starting value of 1.9% (Tucker and Rahilly, 1982): C_0 =1.9, C_1 =2.5, C_2 =−0.04014, and C_3 =0.0002668. Note that not only was C_0 changed but C_1 to C_3 had to be adjusted to suit the revised data. The constants C_0 to

C_3 have been subjected to further revision over time, but all equations developed have final payment at about 117% of practical completion.

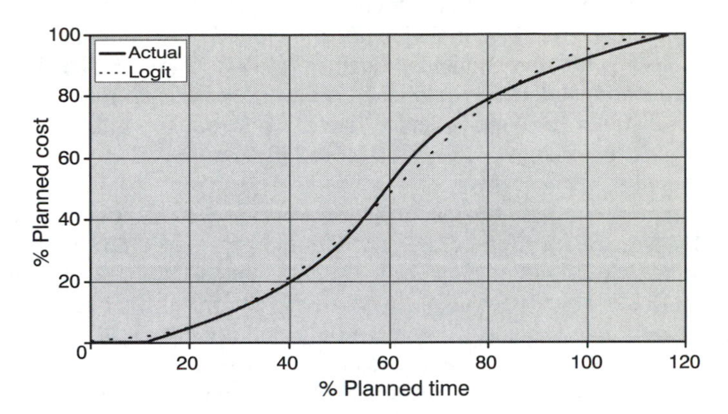

Figure 3.16 The equivalent Logit curve to Bromilow's S curve using Balkau (1975) constants

Figure 3.17 illustrates the relationship expressed using the constants identified by Tucker and Rahilly (1982), as well as the Logit approximation, with $\alpha = 0.0533$ and $\beta = 1.6826$ and SDY= 2.23.

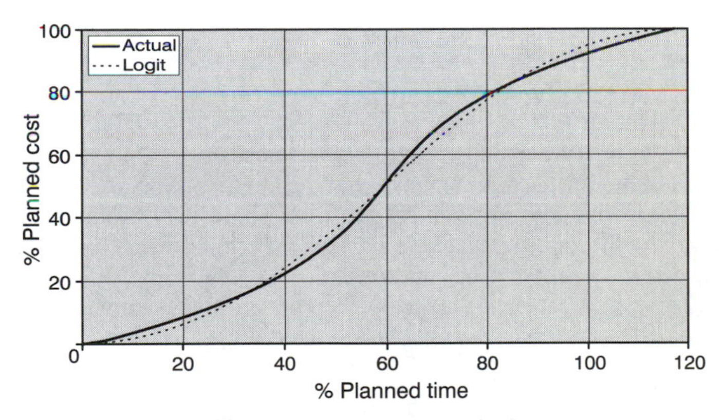

Figure 3.17 The equivalent Logit curve to Bromilow's S curve using Tucker and Rahilly (1982) constants

The Bromilow model is a client model, and as such is aimed at satisfying the needs of client bodies who need to forecast their cash commitments during the life of a project. This model was developed at a time of relatively stable economic conditions, before contractors became interested in these models. Subsequently practical refinements have been incorporated, such as allowances for retention (Tucker and Rahilly, 1985).

The Peer model

It seems likely that Peer (1982) was influenced by the Bromilow model, as his work is similar, although the equation adopted is not an inverted polynomial. Peer initiated a research project with his students Zoisner (1974) and Berdicevsky (1978). Berdicevsky unfortunately never published his work, a task left to Peer. The model here is referred to as the Peer model reflecting his role in guiding the research. However, it must be recognised that it was in large part developed as a Master of Science project by Berdicevsky.

Peer supported the use of standard curves because, as he pointed out, the cash flow forecast is often required before any detailed time schedule can be given. He briefly examined the cumulative ogives for such factors as labour, materials, general costs, subcontractors, etc. Peer also examined the pattern of total cost and contract sum, but for only one project. He performed a polynomial regression on the data points of four major public housing projects. He tested three equations: a fourth degree polynomial, a tanh, and an error function. The standard deviation was found to be less for the fourth degree polynomial, which therefore yielded a more accurate estimate. It is likely that Berdicevsky's work developed the first polynomial expression for value in terms of time. The equations identified for housing and public buildings were:

$$y = 0.0009 + 0.2731t - 1.0584t^2 + 5.4643t^3 - 3.6778t^4 \quad (3.20)$$

$$y = 0.567(\tanh(3.2495t - 2.038) + 0.963) \quad (3.21)$$

$$y = 0.5487(\mathrm{erf}(2.8822t - 1.7972) + 0.986) \quad (3.22)$$

where t = ratio of total time and y = ratio of total cost.

The error function used is unknown, but the 4th degree polynomial and the tanh function curves derive almost identical profiles. These may be matched using the Logit equation, with results tabulated in Table 3.7. Peer noted that his model fitted the data with an error of 2.40% for the entire sample and 1.98, 1.32, 2.76 and 1.68 respectively. The Logit model has an SDY of 0.53% which in these circumstances would provide an excellent match to the underlying data.

Table 3.7 Logit equivalents to Peer's housing project profiles

Peer's equation	α	β	SDY
$y = 0.0009 + 0.2731t - 1.0584t^2 + 5.4643t^3 - 3.6778t^4$	–0.7281	1.6814	0.53%
$y = 0.567(\tanh(3.2495t - 2.038) + 0.963)$	–0.7371	1.7153	0.44%

Berdicevsky (1978) analysed the data from four projects constructed within a university to derive a third-degree polynomial. The validity of this regression for developing standard curves might be questioned as the projects were all '...different in size, design, construction time and intended use' (Berdicevsky, 1978). Berdicevsky's simplified equation was:

$$y = 2.4t^2 - 1.4t^3 \tag{3.23}$$

which Peer (1982) later cited as:

$$y = 0.00101 - 0.00459t + 2.36949t^2 - 1.39030t^3 \tag{3.24}$$

This, and their equation for a combined data set of seven projects:

$$y = 0.0089 - 0.2698t + 2.7909t^2 - 1.5181t^3 \tag{3.25}$$

used up to six significant figures, which seems an unjustified level of precision for the model. Figure 3.18 illustrates these three curves, also Table 3.8.

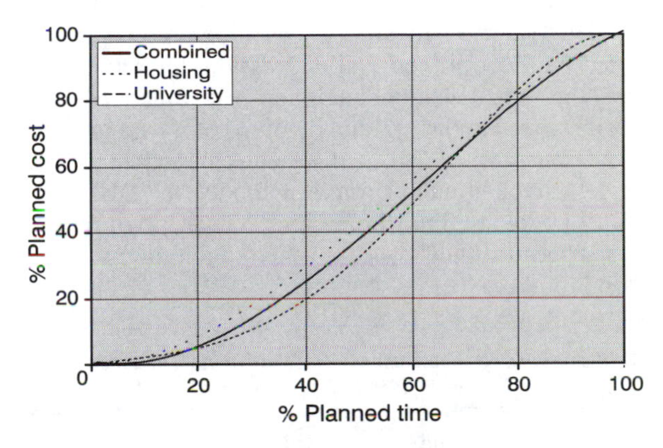

Figure 3.18 Peer's three cost-flow curves

The lack of accuracy can be highlighted by the deviation of the model from projects against which it was tested. Also Berdicevsky tested his model on another project and found it deviated by some seven percent (presumably using a measure equivalent to the SDY measure).

> Considering the dynamic character of a company's cash flow plan, and the need for regular updating, the deviations obtained by this procedure are insignificant for practical use. It should be remembered that management is mainly concerned with the overall cash flow of the firm, rather than that of a single project. It may also be assumed that, when working on several projects in parallel at different stages of progress, deviations of individual projects tend to balance out (Berdicevsky, 1978).

Whereas Bromilow was concerned with cash flows which affect the client, Berdicevsky and Peer were more concerned with gross inward cash flows for the contractor.

Table 3.8 Logit equivalents to Berdicevsky and Peer two-project samples

Peer's equation	α	β	SDY
Four housing projects			
$y = 0.0009 + 0.2731t - 1.0584t^2 + 5.4643t^3 - 3.6778t^4$	–0.7281	1.6814	0.53%
Three university buildings			
$y = 0.00101 - 0.00459t + 2.36949t^2 - 1.39030t^3$	–0.3834	1.4164	1.11%
Combined data set of seven			
$y = 0.0089 - 0.2698t + 2.7909t^2 - 1.5181t^3$	–0.5383	1.4205	1.18%

The Betts and Gunner variation on Peer's model

Betts and Gunner (1993) carried out an analysis of 73 projects from the Pacific Rim. Their work adopted the Peer polynomial expression, although this source was not cited and it is likely they had developed their approach independently. Their methodology was simple, if somewhat flawed, and involved performing a polynomial regression on sets of data for projects from different categories. They introduced and rejected other models, including the Bromilow model, the DHSS model, the Tucker 'Single Alternative' model and the Kenley and Wilson Logit model. These models were rejected either because they 'were not favourable to Pacific Rim projects' or 'were evaluated and found to be inappropriate'.

Table 3.9 Betts and Gunner (1993) polynomial parameters with fitted Logit parameters

Project type	1st	2nd	3rd	α	β	SDY
Overall	0.4013	0.01066	–6.1116E-5	–0.0613	1.1807	1.15%
Main contract	0.1842	0.01638	–9.4801E-5	–0.0598	1.3049	1.61%
Commercial	0.1492	0.01660	–9.4326E-5	–0.1201	1.2830	1.79%
Residential		0.02089	–1.2132E-4	–0.1052	1.3569	2.44%
Industrial	0.4343	0.01244	–8.0063E-5	0.1103	1.2222	1.77%
Institutional		0.01719	–8.8577E-5	–0.3655	1.2526	2.11%
School	0.4328	0.00722	–2.6761E-5	–0.0661	1.3904	2.81%

Of interest was their rejection of the Kenley and Wilson model. They found computing the parameters rather complicated and believed the model to 'be an idiographic post hoc analysis approach which could only be used during construction to predict the cash profile of an individual project'. While the former is true, the latter is only partly. The Logit model is flexible enough to fit most profiles, however the idiographic epistemology of the model challenges the assumptions upon which such nomothetic processes as producing average curves are based. Betts and Gunner desired a model which, like the Bromilow model, was based on planned practical completion and value. In fact their data plots indicate that only 18 of their 73 projects reached 100% value. This logic is flawed however, because their model must forecast not only the cash flow profile, but the cost over- or under-run as well as the time over-run. Most researchers have recognised that it is necessary to separate these two factors if trying to develop standard curves. It is this property of obtaining standard curves which run over or under budget at the end of the project which causes the lower SDY (*n* degrees of freedom) values in Table 3.9.

An examination of the curves upon which the Betts and Gunner model is based reveals that the range of individual projects, about the fitted regression line, is extensive. For example, for commercial projects at $t = 70\%$, actual projects from a sample of only ten vary from $v = 36\%$ to 80%. But Betts and Gunner feel at liberty to conclude that commercial projects are $v = 60\%$ at $t = 70\%$ (which none of the sample of projects were). Given the limited degrees of freedom involved, such a contention is statistically extraordinary. Furthermore, they provide a chart with upper and lower bounds for a forecast—for all projects. This provides a range of $v = 54\%$ to 64% at $t = 70\%$. An examination of the source data indicates that only two of the sample of ten commercial projects actually lie within this range. The discrepancy between the standard profiles, the upper and lower ranges and the individual project profiles (shaded but highlighted by the outer bounds) is clearly seen in Figure 3.19.

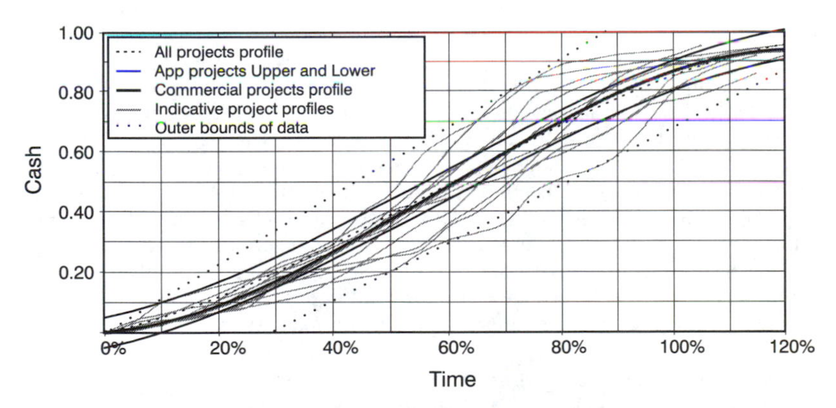

Figure 3.19 Adapted from Betts and Gunner (1993) commercial projects with standard profiles, upper and lower ranges for all projects and outer bounds

Such a conclusion is convenient and not without precedent, so it is unfair to single them out, especially as they did at least graphically show the individual projects upon which the model is based. It is worth providing the Betts and Gunner equation parameters. Their equation is a third degree polynomial of the same form as the Peer (1982) model, but without the intercept term, thus ensuring all projects commence at (0,0). Their parameters are provide in Table 3.9.

A note of caution: this model also has practical problems, for example the *Overall* type has parameters which result in a curve asymptotic to $v = 98\%$ or lower.

Despite the epistemological, methodological and practical problems, the Betts and Gunner model does provide a standard curve. However, rather than being a set of equations which forecast Pacific Rim projects, this model is at best an average of a very small and specific sample and possibly a misleading interpretation of the underlying data.

The Betts and Gunner model, published in 1993, represents the last attempt to explain cash flows by generating models which combine the standard profile with the forecast of end date and value. Subsequent models have accepted the practice, introduced by Kenley and Wilson, of defining the end point *(V,T)* as (100%,100%).

The Tucker compound Weibull–Linear formula

Dr Selwyn Tucker has been a major player in the cash flow arena, and initially worked from the Bromilow S-curve model, updating the constants as part of the effort to develop a new computer software package called FINCASH. Subsequent work identified limitations with this approach, and a new model was developed using a compound formula. The compound formula represented a new step in the epistemology of cash flow forecasting[1].

Tucker adopted a probability approach, using reliability theory. He held that a project could be considered as a population of expenditure payment units, all of which have a similar characteristic, and every one is an amount of money to be paid on condition of completed work. Payment units move from a committed state to a paid state during the project—as the work is completed.

> If it is assumed that the characteristics of the payment units are all similar and that they are treated under similar conditions, the payment unit completion time T may be considered as having a probability distribution. This is analogous to a population of items, such as a bundle of banknotes put into circulation on the same day, completing their intended function by being lost or destroyed. The probability that a particular banknote selected at random has served its intended purpose and disappeared from circulation at time t can be described in terms of reliability theory ...The following development of cash flow distributions draws heavily upon this theory by assuming the change in state of payment units from 'committed' to 'paid' as being equivalent to failure of an item (Tucker, 1986).

While it could be argued that the logic is obscure (having the same probability of payment as the probability of loss or destruction of banknotes sounds like a recipe for never getting paid) the mathematics is clear. The S-curve is created by combining a linear distribution and a Weibull distribution. This reflects the fixed cost components combined with the distributable components.

The cumulative S curve represents the cumulative probability of expenditure $P(t)$. The two component probability (periodic) curves $p(t)$ are combined with a weighting (k and k') where the total weighting equals unity. The components, given in equation 3.26 and charted in Figure 3.20, are:

---

[1] FINCASH is a computer software package developed by the CSIRO and remains available to this day. The CSIRO has continued to develop this package over time.

if $k+k'=1$ then $k'=1-k$

$$p_1(t)=k\left((b/c)(t/c)^{b-1}\,e^{\left(-(t/c)^b\right)}\right) \text{ Weibull distribution}$$

$$p_2(t)=(1-k)a \text{ Linear distribution}$$

$$\text{and } p(t)=k\left((b/c)(t/c)^{b-1}\,e^{\left(-(t/c)^b\right)}\right)+(1-k)a \text{ Combined distribution} \quad (3.26)$$

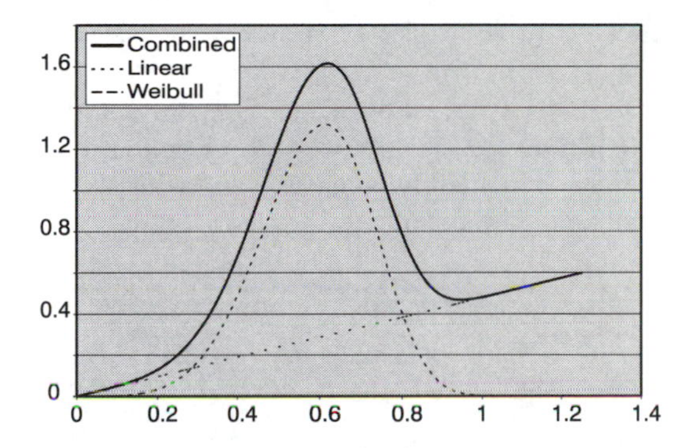

Figure 3.20 Tucker's component probability distributions ($p(t)$)

The cumulative distribution $P(t)$ is similarly provided by the following Equation 3.27, and is charted in Figure 3.21:

if $k+k'=1$ then $k'=1-k$

$$P_1(t)=k\left(1-e^{\left(-(t/c)^b\right)}\right) \text{ Weibull distribution}$$

$$P_2(t)=(1-k)at \text{ Linear distribution}$$

$$\text{and } P(t)=k\left(1-e^{\left(-(t/c)^b\right)}\right)+(1-k)at \text{ Combined cumulative distribution} \quad (3.27)$$

Tucker aimed to produce an alternative to the Bromilow S-curve model, which he considered '...to be an excellent approximation to the empirical form' (Tucker 1986). Figure 3.22 illustrates the new Tucker model compared with Bromilow's model. Also shown is the Logit equivalent of the Tucker curve. The Logit model parameters are in Table 3.10.

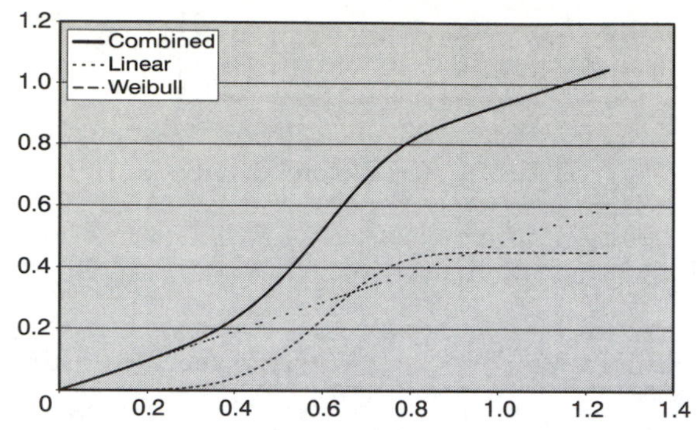

Figure 3.21 Tucker's component cumulative distributions $P(t)$

The SDY for Tucker's 1986 model when compared with Bromilow's model using Tucker and Rahilly (1982) is 2.04%. It is clear that each of these models achieves a profile which is remarkably consistent and with only an SDY of approximately 2% when compared with the others.

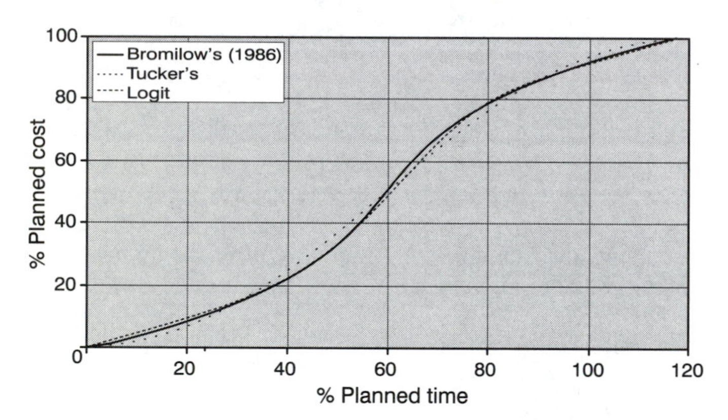

Figure 3.22 Tucker's S-curve compared with Bromilow's constants (Tucker and Rahilly, 1982) and the Logit model

The combination of the linear distribution and the Weibull distribution does have significant intuitive attraction. Logically, it is likely that projects do in fact consist of a linear component (being the fixed costs) and a variable component—that which forms the S shape.

Table 3.10 Logit equivalents to Tucker's models

Model	α	β	SDY
Bromilow's model using Tucker and Rahilly (1982)	–0.0533	1.6826	2.23%
Tucker's model using Tucker (1986)	–0.0312	1.6277	2.25%

This model was tested further in an attempt to establish a single alternative formula in 1988. This will be discussed later, but is mentioned now because the model it was tested against was the DHSS model.

The DHSS model

A most significant work in the study of construction project gross cash flows was started at the British Department of Health and Social Security (DHSS) by Hudson in 1967, and was published by Hudson and Maunick (1974) and again by Hudson (1978). This model was the first to contain only two constants in the gross cash flow equation and to provide a system of standard curves for sub-categories of projects. The DHSS model was aimed at the requirements of a large client body which required improved standards of capital investment control in much the same way as the Bromilow model.

The DHSS cash flow has wide acceptance amongst quantity surveyors, particularly those influenced by the UK through their education, it is thus an important model. Its approach is similar to the Bromilow model in that the derivation of the polynomial curve was intended to provide best-fit to the available data. However, their approach was quite different, in that:

- They expressed value in terms of time (generally considered a fault of the Bromilow model).
- They argued that all projects have properties of 'rate of expenditure' in common. (This was not stated as such, but rather evidenced by their treatment of fixed points of expenditure common to all hospital projects.)
- They expressed both axes on a scale to a maximum of ratio 1.0 (100%).

The DHSS model holds that all projects commenced at a similar rate, and therefore after twelve months had completed £0.9 million and similarly after twenty–four months had expended £2.1 million.

Establishing these two fixed points had the effect of determining the shape of any projects' cash flow profile, by using those two points in their forecasting formula to establish two simultaneous equations. The model then requires the user to use real points to recalculate the profile.

The Hudson model of project cash flows is an analytical model capable of achieving standard curves for categories of project, and also of adapting to each individual project using real data points and simultaneous equations.

The equation used for the sigmoid curves was:

$$y=\frac{v}{S}=x+cx^2-cx-\frac{(6x^3-9x^2+3x)}{k} \qquad (3.28)$$

where v is the cumulative monthly value of work executed before deduction of retention moneys or addition of fluctuations, S is the contract sum (or 1.0 for standard curves) and c and k are two parameters sufficient to define the curve. Figure 3.23 illustrates the DHSS curve compared with an equivalent Logit curve.

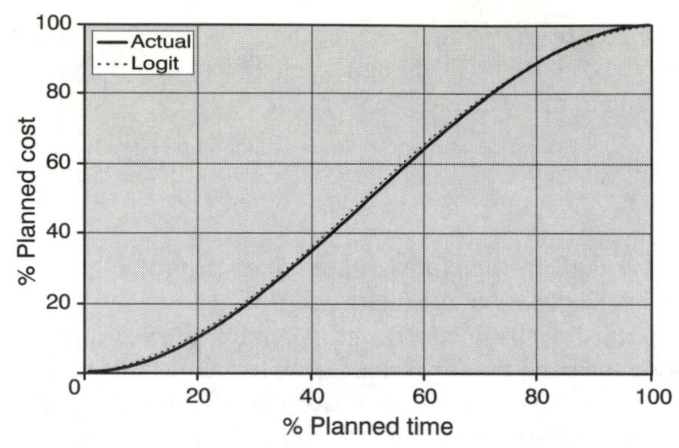

Figure 3.23 DHSS standard curve parameters (Hudson, 1978) with Logit equivalents, Value—to £30k; DHSS: $c = -0.041$, $k = 7.018$; Logit: $\alpha = 0.045$, $\beta = 1.203$

Figure 3.24 illustrates the DHSS curve compared for three different project value parameters.

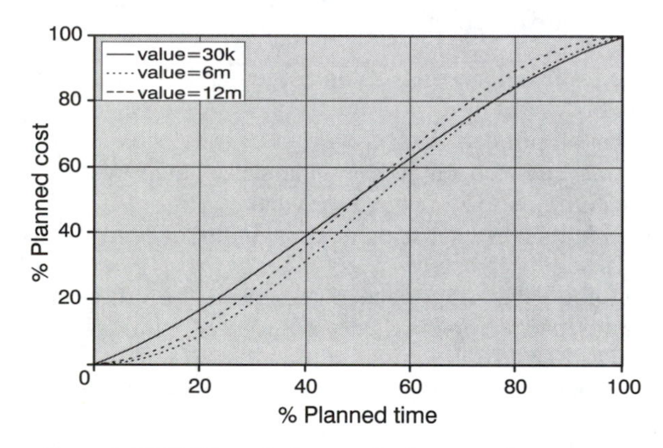

Figure 3.24 DHSS standard curves for three sets of parameters

It was found to be possible to derive values of c and k for different categories of work, however:

> It will be appreciated that as the lines are curves of best fit, no one job will necessarily fit the curve exactly. However, when all jobs in a cost category are taken cumulatively, the forecast produces a close approximation to the actual value of work executed for schemes in that category (Hudson and Maunick, 1974).

The basic workings of the model were as follows. First the cash flow could be expressed by Equation 3.28 where both axes were expressed as ratios of final amounts. Thus the scales ranged from 0 to 1. The relationship for a group of projects could be found by a polynomial regression of all the data points for the projects. (This is similar to the Bromilow method.) Then by taking any two points on the curve (fitted curve), reading off the x and y coordinates of these points and substituting them into the forecasting formula, two simultaneous equations could be generated and solved for c and k. These then are the 'standard parameters' and the two variables which enable the model to adapt its shape. This is a post-hoc analysis requiring the final contract sum and period to be known for each project. These figures were taken by Hudson to be at practical completion.

Secondly the model was used for prediction. Using the values of c and k derived for a relevant job category and the project cost, a likely works duration could be calculated. In turn, a prediction of cash flows at specified time intervals could be made.

Table 3.11 DHSS standard curve parameters (Hudson, 1978) with Logit equivalents

Cost cat £	c	k	α	β	SDY
10k to 30k	(0.041)	7.018	0.045	1.203	0.18%
30k to 75k	(0.360)	5.000	0.410	1.314	0.89%
75k to 120k	(0.240)	4.932	0.273	1.308	0.59%
120k to 300k	(0.200)	4.058	0.231	1.377	0.58%
300k to 1.2m	(0.074)	3.200	0.087	1.494	0.34%
2	0.010	4.000	0.012	1.382	0.06%
3	0.110	3.980	(0.128)	1.386	0.32%
4	0.159	3.780	(0.184)	1.409	0.49%
5	0.056	3.323	(0.066)	1.473	0.25%
6	0.192	3.458	(0.225)	1.453	0.65%
6.5	0.154	3.401	(0.180)	1.459	0.54%
7	0.172	3.557	(0.202)	1.440	0.57%
7.5	0.131	3.445	(0.154)	1.455	0.46%
8	0.142	3.538	(0.166)	1.440	0.47%
8.5	0.099	3.404	(0.116)	1.459	0.36%
9	0.104	3.456	(0.123)	1.454	0.37%
9.5	0.061	3.317	(0.072)	1.474	0.27%
10	0.063	3.344	(0.074)	1.470	0.27%
10.5	0.019	3.207	(0.022)	1.493	0.22%
11	0.018	3.218	(0.210)	1.491	0.21%
11.5	(0.025)	3.089	0.030	1.515	0.27%
12	(0.028)	3.090	0.033	1.515	0.28%

Hudson developed a technique for applying the model to individual projects, using data from the project itself to generate the function. As only two data points were used to form the simultaneous equations, either two actual points from the data had to be selected, or a regression performed on the available data and two points chosen from the resultant fitted curve. These points would then provide the constants for the simultaneous equations from which c and k values for the project (as an estimate from available data) could be derived. This could be further revised as the project proceeded. The 'smoothing' regression is used because '…the flow of expenditure on a scheme does not follow a regular smooth curve' (Hudson, 1978).

Hudson and Maunick published a table of standard c and k values for differing project values.

These are outlined in Table 3.11 together with the equivalent α and β values for each curve and the goodness of fit (SDY). The calculated SDY values indicate that the Logit model is in fact a very good approximation for the Hudson DHSS model.

The DHSS curves become much clearer when viewed in real value and time rather than the normalised ratios (Figure 3.25). This reflects the true value and contribution of Hudson and Maunick, which is the identification of a path that all projects follow—allowing both the forecasting of cash flow profiles as well as the duration of projects. Of course this assumption is not necessarily correct, and was only claimed for hospital projects. It could only be valid at the time of the study and would be subject to price adjustment over time.

This reliance on real data illustrates a clear problem with the use of the DHSS model for forecasting, as it must necessarily be subject to reassessment, and regularly updated for inflation (Berny and Howes, 1982).

The greatest problem, however, is the assumption that the rate of work for the first £0.9 million is constant and the first £3 million is constant for a project value above £6 million which seems unlikely to hold true, as larger projects would be able to spend more money in a 12 or 24 month period.

There is often confusion regarding forecasting, with some authors indicating that a match to the data equates to a good forecast. While the DHSS model may be a good match to its originating data, its value as a forecasting model for future projects has been challenged. Sidwell and Rumball (1982) demonstrated that the model failed in forecasting, with 30 instances in a sample of 38 projects where the DHSS model could not be shown to be a reasonable predictor.

Berny and Howes

Berny and Howes (1982) recognised several shortcomings in the Hudson model and followed up with some modifications. They re-expressed the Hudson equation 3.28 into a form which more easily allowed for manipulation. Their work contributes to the understanding of the mathematics of the sigmoid, but unfortunately does not really solve the fundamental problems which they found to be inherent in the Hudson model. They were particularly concerned with over-run, which is met by including peripheral mechanisms, such as segmented curves, which, it can be argued, make the model unnecessarily complex.

The Berny and Howes model (1982) had its roots within the Hudson model. It is also an analytical model capable of encompassing both standard curves and individual project curves. The authors' stated aim was to overcome shortfalls which had been identified in the Hudson model. The Berny and Howes model is an improvement on the Hudson model and this was achieved in two ways. First, a new mathematical model was derived, re-expressing the Hudson equation. Secondly, peripheral mechanisms were attached to cope with other problems.

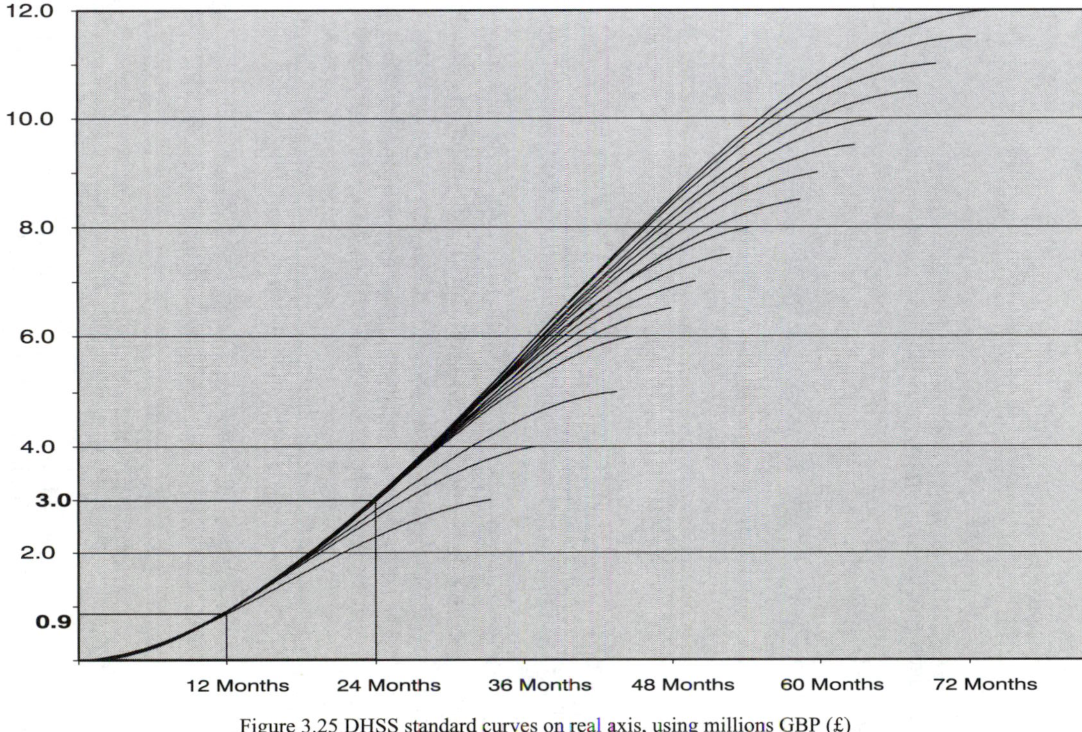

Figure 3.25 DHSS standard curves on real axis, using millions GBP (£)
(Re-calculated using the DHSS formula, to replicate Hudson, 1978: Figure 2)

The Berny and Howes model, at the basic level, is a re-expression of the Hudson model given in Equation 3.28. Their equation is:

$$y=\frac{v}{S}=x(1+a(1-x)(x-b)) \tag{3.29}$$

where a and b are constants.

This can be re-expressed:

$$y=-ax^3+(a+ab)x^2+(1-ab)x \tag{3.30}$$

and if the Hudson model is also re-expressed:

$$y=-6x^3+(c+9)x^2+(1-c-3)x \tag{3.31}$$

then under the following conditions the two models are equivalent:

$$c=ab-a/2$$

$$k=6/a \tag{3.32}$$

or

$$b=(ck+3)/6 \equiv (0.5+ck/6)$$

$$a=6/k \tag{3.33}$$

However, the Berny and Howes model does allow for the parameter b to be calculated from the shape of the data, as the point of inflection (peak expenditure) is at x_p, where:

$$x_p=(b+1)/3 \tag{3.34}$$

$$\text{therefore } b=3x_p-1 \tag{3.35}$$

The calculation of a is more problematic. Berny and Howes found that good results are given by:

$$a=\min\left\{\left(\frac{1}{1.21-1.1b}\right),\left(\frac{1}{1.1b+0.11}\right)\right\} \tag{3.36}$$

Alternatively they recommended computer generation of a family of 'S' shapes with the best fit being selected visually. Finally they suggested deriving the estimated peak period cost (Y_p) where:

$$Y_p=1+a(3x_p(x_p-1)+1) \tag{3.37}$$

Unfortunately this re-expression did not solve the problem of over-run, which led Berny and Howes to incorporate a supplementary exponential equation using the parameter gamma. This was included in a computer software package, but the formula was not published, nor was the method of interaction apart from noting that the point of change from one equation to the other was derived iteratively.

The inclusion of the exponential equation improved SDY from a maximum of 15% to below 5%. It would be interesting to compare their results with the results of the Logit cash flow model on similar data; however neither Hudson (for the DHSS model) nor Berny and Howes (for their model) have published source data for this comparison.

It is possible to use this model for the generation of standard curves or for individual projects, much the same as the DHSS model. The equivalent *a* and *b* values to the DHSS standard parameters are given in Table 3.12.

Table 3.12 Berny and Howes (1982) standard curve parameters with Logit equivalents

Cost cat £	C	K	a	b	α	β	SDY
10k—30k	–0.041	7.018	0.855	0.452	0.045	1.203	0.18%
30k—75k	–0.360	5.000	1.200	0.200	0.410	1.314	0.89%
75k—120k	–0.240	4.932	1.217	0.303	0.273	1.308	0.59%
120k—300k	–0.200	4.058	1.479	0.365	0.231	1.377	0.58%
300k—1.2m	–0.074	3.200	1.875	0.461	0.087	1.494	0.34%
2	0.010	4.000	1.500	0.507	0.012	1.382	0.06%
3	0.110	3.980	1.508	0.573	–0.128	1.386	0.32%
4	0.159	3.780	1.587	0.600	–0.184	1.409	0.49%
5	0.056	3.323	1.806	0.531	–0.066	1.473	0.25%
6	0.192	3.458	1.735	0.611	–0.225	1.453	0.65%
6.5	0.154	3.401	1.764	0.587	–0.180	1.459	0.54%
7	0.172	3.557	1.687	0.602	–0.202	1.440	0.57%
7.5	0.131	3.445	1.742	0.575	–0.154	1.455	0.46%
8	0.142	3.538	1.696	0.584	–0.166	1.440	0.47%
8.5	0.099	3.404	1.763	0.556	–0.116	1.459	0.36%
9	0.104	3.456	1.736	0.560	–0.123	1.454	0.37%
9.5	0.061	3.317	1.809	0.534	–0.072	1.474	0.27%
10	0.063	3.344	1.794	0.535	–0.074	1.470	0.27%
10.5	0.019	3.207	1.871	0.510	–0.022	1.493	0.22%
11	0.018	3.218	1.865	0.510	–0.210	1.491	0.21%
11.5	–0.025	3.089	1.942	0.487	0.030	1.515	0.27%
12	–0.028	3.090	1.942	0.486	0.033	1.515	0.28%

Tucker's single alternative Weibull formula

As mentioned previously, Tucker further developed his Weibull model as a single alternative model for the DHSS curves (Tucker 1988). With the benefit of hindsight, Tucker re-expressed his 1986 Weibull equation, which was:

$$P_1(t) = k(1 - e^{(-(t/c)^b)}) \quad \text{(from 3.27)}$$

$$\text{giving } Y_w = \alpha\left(1 - e^{(-((x-\delta)/y)^\beta)}\right),\ x > \delta \text{ or } Y_w = 0,\ x \le \delta \quad (3.38)$$

thereby retrospectively introducing a new element δ which is a time-offset factor, α is a new normalisation constant equivalent to S in the DHSS formula.

Tucker's single alternative model for the DHSS removed the linear component identified in the 1986 compound formula. This is a shame, and perhaps it would have been better left in, with the value of a set to 0.0, as follows:

$$Y_w = \alpha\left(k\left(1 - e^{(-((x-\delta)/y)^\beta)}\right) + (1-k)ax\right),\ x > \delta \text{ or } \alpha\left(Y_w = (1-k)ax\right),\ x \le \delta$$

$$\text{where } a = 0.0 \quad (3.39)$$

but in the particular circumstances, this would be understandably confusing, and redundant as 'visual inspection of the DHSS curves did not show the characteristic which necessitated its inclusion for the Bromilow curve' (Tucker 1988). Tucker's subsequent manipulation of the formula to normalise for the effect of δ results in the following equation:

$$Y'_w = 1 - e^{(-((x-\delta)/\gamma)^\beta + (-\delta/\gamma)^\beta)} \quad (3.40)$$

Tucker then compared this model with the DHSS constants for various dollar ranges. These are shown in Table 3.13 together with the Logit values found earlier. This shows that the Tucker alternative model fails to achieve the same degree of fit as the Logit model.

Tucker noted that the new model 'retained two important features of the DHSS model, (i) independent changes to the shape or spread of the curves can be achieved by varying only one parameter, and (ii) estimation of the likely duration of the project is possible from knowledge of one point on the curve'. (Tucker, 1988).

Sing and Phua

Sing and Phua (1984) did not develop a standard curve for cash flow profiles. They are mentioned here because their work clearly demonstrated the wide range of profiles possible for cash flow S curves. Their work harkened back to Bromilow's original work, in that they defined a curve envelope. As with Bromilow's confidence limits, they realised that the size of curve envelope that covered cash flow profiles with only a 50% confidence was enormous, indeed occupying nearly 50% of the available space on the chart.

Table 3.13 Tucker (1988) Single alternative formula compared with Logit equivalents

Cost cat £ m	β (Shape)	γ (Scale)	δ	χ^2	α **Logit**	β **Logit**	SDY
10k—30k	10.02	3.71	−3.293	0.48	0.045	1.203	0.18%
30k—75k	8.70	2.79	−2.320	0.46	0.410	1.314	0.89%
75k—120k	8.47	2.68	−2.183	0.37	0.273	1.308	0.59%
120k—300k	7.83	2.27	−1.746	0.39	0.231	1.377	0.58%
300k—1.2m	6.65	1.73	−1.167	0.48	0.087	1.494	0.34%
2.0	5.44	1.50	−0.924	0.45	0.012	1.382	0.06%
3.0	5.16	1.40	−0.797	0.48	−0.128	1.386	0.32%
4.0	4.99	1.28	−0.673	0.98	−0.184	1.409	0.49%
5.0	5.07	1.29	−0.699	0.70	−0.066	1.473	0.25%
6.0	4.87	1.23	−0.615	0.92	−0.225	1.453	0.65%
6.5	4.80	1.19	−0.579	1.19	−0.180	1.459	0.54%
7.0	4.74	1.19	−0.576	1.09	−0.202	1.440	0.57%
7.5	4.75	1.21	−0.603	0.71	−0.154	1.455	0.46%
8.0	4.67	1.20	−0.592	0.73	−0.166	1.440	0.47%
8.5	4.68	1.19	−0.592	0.68	−0.116	1.459	0.36%
9.0	4.66	1.19	−0.591	0.67	−0.123	1.454	0.37%
9.5	5.06	1.28	−0.692	0.69	−0.072	1.474	0.27%
10.0	5.23	1.33	−0.739	0.69	−0.074	1.470	0.27%
10.5	5.23	1.31	−0.729	0.71	−0.022	1.493	0.22%
11.0	5.22	1.31	−0.729	0.67	−0.210	1.491	0.21%
11.5	5.28	1.30	−0.732	0.72	0.030	1.515	0.27%
12.0	5.40	1.34	−0.765	0.74	0.033	1.515	0.28%

They also were the first to break cash flow profiles into components for different trades, and show how a project was made up from a whole series of mini S curves for each trade, all combining to make one project curve. This goes a long way to explain why projects can be so variable in their profiles.

Khosrowshahi

Khosrowshahi (1991) introduces a different way of looking at the cash flow profile. He introduces some heuristics which a control profile must follow—and develops a conforming mathematical model. He then adds two additional components to account for real profiles: distortion and kurtosis modules. He justifies this approach by arguing that other methods are flawed by various shortcomings. He argues:

> The traditional method of project expenditure budgeting is time consuming, costly and subjective to cumulative error. Furthermore, it is a prerequisite that the project programme must be established before a budget can be determined. The alternative approach, as well as having a limited range of application and lacking flexibility, often provides no means for the user to comprehend the logic of prediction, hence being alienating to the user. In instances where the user is required to provide input, an extensive knowledge, experience and information is demanded from the user (Khosrowshahi, 1991).

Khosrowshahi attempts therefore to develop a model which is tied to the shape of the curve in a way that can be used and understood by an estimator.

The model concentrates on the periodic expenditure rather than the cumulative expenditure, as it is here that changes can be seen in the actual cash flow. The user concentrates on the critical point of inflection of the cumulative curve, or the point of peak expenditure of the periodic expenditure. This point is easily visually determined, and occurs at (Q,R), where $Q = Y_p / C$ and $R = X_p / n$. This is based on real values (and for time, is based on a fixed number of n periods). When ratio or percentage data is used, C=1.0 (100%), $Q = Y_p$ and $R = X_p$.

There are four general characteristics of the periodic curve, which according to Khosrowshahi are common to all projects. These are:

1. *Non-presence of negative values.*
2. *Periodic values are discrete*[2].
3. *Initial and end values are zero.*
4. *The periodic curves have a two phased monotonic nature*—The first phase is a growth phase and the second phase is a decay phase.

Further, there are four specific characteristics of the periodic curve, which according to Khosrowshahi, separate individual projects. These are:

1. The coordinates of the peak period on the graph representing the periodic expenditure versus time.
2. The position and intensity of any possible distortions which cause troughs and minor peaks on the periodic expenditure pattern.
3. The measure of the overall rate of growth commencing from time zero to the peak point.
4. The measure of the overall rate of decay commencing from the peak point to the end of the project.

These characteristics were used as rules which the model was intended to meet, the general characteristics through the control module and the specific characteristics through the distortion and kurtosis modules.

It is the control module which is of primary concern here as this relates to the generic project and standard forms of cash flow. Khosrowshahi did not publish the

[2] There is some doubt about the validity of this position. Expenditure at any point in time is not independent of previous expenditure and in fact is to some degree determined by the current rate of expenditure. Thus a discrete 'low' periodic expenditure is less likely during a period of 'high' periodic expenditure. This is highlighted by characteristic 4.

formula for the other modules, as they were calculated for each project. It is unclear how they were used in the cash-flow program TASC which was the product of his work.

The Khosrowshahi generic cash flow formula for periodic cash flows is given by Equation 3.41:

$$y = C(e^{ax^b(1-x)^d} - 1) \tag{3.41}$$

The derivation of this equation is complex, and relates to an attempt to adjust a basic Weibull probability distribution (as the closest match) to meet the four rules for generic cash flows. More details can be found in Khosrowshahi (1991).

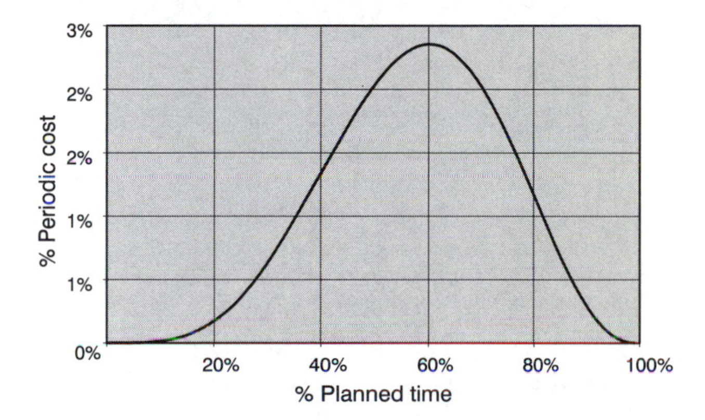

Figure 3.26 Khosrowshahi's periodic distribution where $Q = 0.15$, $R = 0.45$ and $d = 2.6955$

The key to this model is that the constants may be related back to the point (Q,R). Khosrowshahi derived that a and b may be calculated from Q, R and d:

$$a = \frac{dR}{(1-R)} \tag{3.42}$$

$$b = \frac{\log_e(1+Q)}{R^a(1-R)^d} \tag{3.43}$$

with d being iteratively adjusted to achieve a cumulative total equal to C. This may be done in practice, but is not simple and requires user interaction to adjust the value of d[3]. The equations represent the periodic distribution. Figure 3.26 shows the periodic distribution where $Q = 0.15$, $R = 0.45$ and $d = 2.6955$ (derived iteratively).

[3] Those using Microsoft Excel should note that the function Ln must be used for Log_e not Log. However, if using Visual Basic functions, then Log is defined as the natural logarithm and is the correct function to use.

The Khosrowshahi model can be shown as a cumulative distribution (by simply summing the periodic flows) and, consistent with the other models, may be compared with a Logit equivalent. Figure 3.27 shows the Khosrowshahi's equivalent cumulative distribution for Figure 3.26, where $Q = 0.15$, $R = 0.45$ and d =2.6955, with the Logit equivalent.

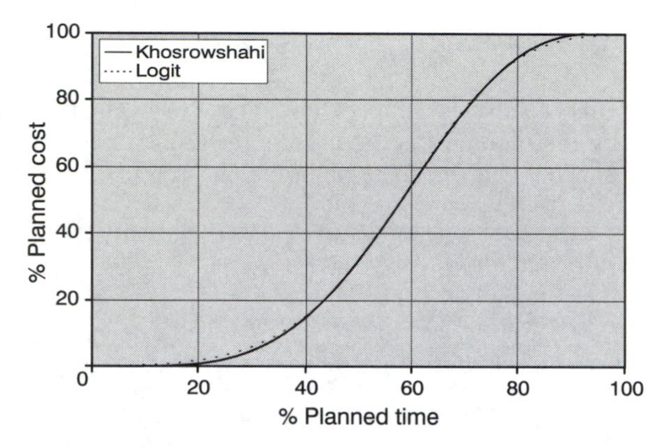

Figure 3.27 Khosrowshahi's cumulative distribution where $Q = 0.15$, $R = 0.45$ and $d = 2.6955$ and the Logit equivalent, where $\alpha = -0.7666$ and $\beta = 2.3686$ and SDY = 0.58%

Khosrowshahi indicates that the required range of Q and R values is broad, and he does not provide any standard profile. However, he did indicate that the average ranges were for Q from 0.04 to 0.22 with mean 0.1 and R from 0.21 to 0.83 with mean 0.496. Taking these mean and extreme values as pseudo 'standard curves' gives the table of parameters, Table 3.14.

Table 3.14 Logit equivalents to Khosrowshahi's mean and extreme values

#	Range position	Q	R	d	α	β	SDY
1	Mean	0.10	0.496	2.774	–0.8656	2.3886	0.56%
2	Min Q, Min R	0.04	0.21	5.795	2.3050	2.1883	1.08%
3	Min Q, Max R	0.04	0.83	0.00215	–0.0156	1.0331	0.29%
4	Max Q, Min R	0.22	0.21	0.048	–0.1749	1.0949	0.23%
5	Max Q, Max R	0.22	0.83	0.00247	–0.2413	1.0741	0.22%

Although Khosrowshahi's stated aim was to make the use of the model simpler and more comprehensible for a user, it is difficult to see that this has in fact been achieved. The addition of the third parameter d, which interacts with both a and b, makes the model far from intuitive. The further addition of the distortion and kurtosis modules makes the model more powerful but even more complex.

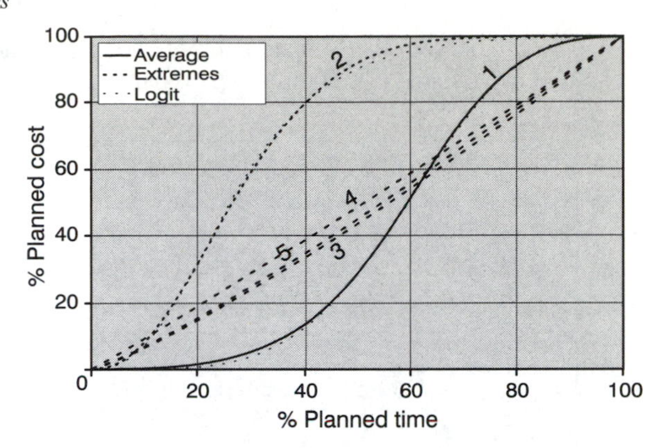

Figure 3.28 Khosrowshahi's ranges of parameters

The Khosrowshahi cumulative distributions which correspond with Table 3.14 are illustrated in Figure 3.28, with equivalent Logit curves.

Miskawi

Miskawi's (1989) model is interesting because, derived from industry practice in the petrochemical industry, it appears to have been developed without reference to existing models. The equation and form are unique:

$$P = \frac{3^t}{2}\sin\left(\frac{\pi(T-t)}{2T}\right)\sin\left(\frac{\pi t}{T}\right)\ln\left(\frac{t+{}^{T}\!/_{2}}{a+t}\right) - \frac{2t^3}{T^2} + \frac{3t^2}{T} \tag{3.44}$$

where P is value, t is the progress time, T is the maximum project duration and a is a shaping factor.

This formula is difficult to work with, and Miskawi simplified it by defining T as 100 and re-expressing the formula. This also has the potential for confusion, and it is easier to form the equation as a ratio where $T = 1.0$ (100%):

$$v = \frac{3^t}{2}\sin\left(\frac{\pi(1-t)}{2}\right)\sin(\pi t)\ln\left(\frac{t+0.5}{a+t}\right) - 2t^3 + 3t^2 \tag{3.45}$$

where v is the progressive percentage of total value, t is the progressive percentage of total time, and a is the shaping constant which, for project S curves, lies between $a = 2\%$ and $a = 97\%$.

Miskawi's formula was assessed for comparison with the DHSS model by Skitmore (1992), who changed the term ${}^{3^t}\!/_{2}$ to ${}^{3^b}\!/_{2}$, as he considered that the component became inappropriate where t approached 100. However, there does not seem to be any such problem with the above formula, with the resultant curves illustrated in

Figure 3.29. A visual comparison of the Miskawi curves shows that they are not the same as those illustrated in Miskawi (1989). This discrepancy cannot be explained.

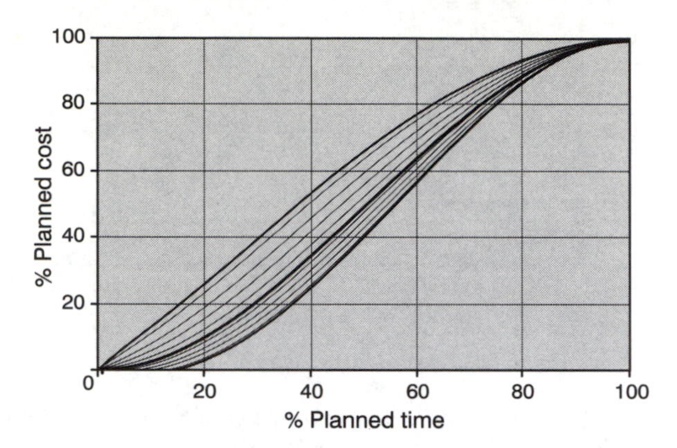

Figure 3.29 Miskawi's S curves for $a = 5\%$ to 95%

The Miskawi curves may be replicated by Logit equivalents. It is interesting to note that the Miskawi Equation 3.44 uses only one variable to describe the S curve, however the range of profiles is very limited. An examination of the equivalent Logit curves suggests that the Miskawi curve is similar, but effectively holding one parameter, β, as a constant and varying α. The Logit α and β equivalents, with SDY values are given in Table 3.15.

Table 3.15 Logit equivalents to Miskawi's range of a values

#	a	α	β	SDY
1	5%	0.7017	1.2623	1.23%
2	10%	0.5951	1.2931	1.04%
3	20%	0.4150	1.3537	0.73%
4	30%	0.2625	1.4295	0.49%
5	40%	0.1243	1.4837	0.31%
6	50%	0.0000	1.5327	0.30%
7	60%	–0.1138	1.5780	0.47%
8	70%	–0.2185	1.6194	0.71%
9	80%	–0.3151	1.6566	0.97%
10	90%	–0.4040	1.6907	1.25%
11	95%	–0.4457	1.7064	1.40%

It is unlikely that these curves will be of much use, as they are too limited in their range of possible shapes and apart from Skitmore's (1992) iterative approach, there is no easy way to derive the parameters from a data set. Indeed, Skitmore discarded the model because he found it was unsuitable for modelling the DHSS curves.

Boussabaine and Elhag

Boussabaine and Elhag (1999) developed a model specifically targeted at the idiographic nature of construction project cash-flow curves, by introducing the fuzzy theory and applying it to cash flow profiles. To do this however, they needed an originating profile, which they derived from an analysis of 30 projects. The model is a compound equation, using Cooke and Jepson's (1979) breakdown of the three stages of the project into (i) parabolic growth for the first third, (ii) linear accumulation for the middle third, and (iii) decaying accumulation for the final third.

if $0 \le x \le \frac{1}{3}$ then $y = \frac{9x^2}{4}$

or if $\frac{1}{3} \le x \le \frac{2}{3}$ then $y = \frac{3x}{2} - \frac{1}{4}$

and if $\frac{2}{3} \le x \le 1$ then $y = \frac{9x}{2} - \frac{9x^2}{4} - \frac{5}{4}$

which together form the Algorithm (3.46)

The resultant S curve is illustrated in Figure 3.30, together with the fitted Logit curve, which has parameters $\alpha = 0.0$ and $\beta = 1.6224$ and SDY=0.52%.

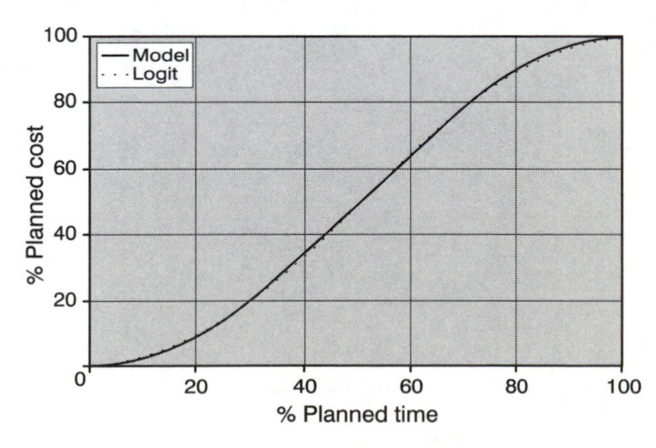

Figure 3.30 Boussabaine and Elhag's (1999) underlying profile for fuzzy analysis and the Logit equivalent, where $\alpha = 0.0$ and $\beta = 1.6224$ and SDY= 0.52%

Experimental approaches

The main contribution of Boussabaine and Elhag lay not in the cash flow profile (which was at the time rather quaint), but rather in their approach, which confronted the idiographic nature of the profiles through the application of fuzzy logic. They believed that this issue had not yet been confronted in the literature (which was not the case, refer Kenley and Wilson, 1986) and thus sought from first principles a strategy to assist management. Fuzzy logic was the method they adopted to manage uncertainty and ambiguity in cash flow forecasts, as it is intended to deal with uncertain or imprecise situations and decisions, which they termed 'fuzzification'.

The fuzzification of the cash-flow profile, substitutes for y the membership function $\mu(x)$ of the cumulative proportion of project cost $(0 \leq x \leq 1)$ is as follows:

if $\quad 0 \leq x \leq 1/3 \quad$ then $\quad \mu(x) = \frac{9x^2}{4}$

or if $\quad 1/3 \leq x \leq 2/3 \quad$ then $\quad \mu(x) = \frac{3x}{2} - 1/4$

and if $\quad 2/3 \leq x \leq 1 \quad$ then $\quad \mu(x) = \frac{9x}{2} - \frac{9x^2}{4} - 5/4$

which together form the Algorithm (3.47)

Their approach was to model three intervals: low cash flow, medium cash flow (the most likely) and high cash flow. These intervals were used to construct the membership function of possible cash flow at any period of a project's progress. The membership being given as follows:

(a) For low cash flow:

$$\mu(x) = \left|\frac{a-x}{b}\right| \quad \text{for } a-b \leq x \leq a \tag{3.48}$$

(b) For medium cash flow:

$$\mu(x) = \left|\frac{x-a+b}{b}\right| \quad \text{for } x \leq a \tag{3.49}$$

$$\mu(x) = \left|\frac{x-a-b}{b}\right| \quad \text{for } x \geq a \tag{3.50}$$

(c) For high cash flow:

$$\mu(x) = \left|\frac{x-a}{b}\right| \quad \text{for } a \leq x \leq a+b \tag{3.51}$$

where a is the mean and b is the standard deviation.

The detailed application of the fuzzy logic model to cash flows is discussed in full in Boussabaine and Elhag (1999). The outcome of the model is the potential to produce a priori modelling of cash flows of unique and complex form, although their application is yet to be demonstrated. They are designed however, for forecasting, and the model appears unsuitable for fitting detailed curves to existing data.

Boussabaine continued with his experimentation, developing a neural networks technique for cash flow forecasting (Boussabaine and Kaka, 1998) and applying it to cost-flow forecasting for water pipeline projects (Boussabaine, Thomas and Elhag, 1999).

The neural networks method targets the assumed inability of regression-based models to handle qualitative factors influencing the profile. It is claimed that neural networks are superior for handling such factors (Boussabaine and Kaka, 1998), although Boussabaine, Thomas and Elhag (1999) note that further testing is required to ascertain whether this approach is more accurate than traditional approaches. Boussabaine and Kaka (1998) did not provide their underlying cash flow model.

It may be concluded here that the fuzzy approach and the neural networks approach are genuine attempts to solve the problem of failure of 'standard' curves for prediction, failure which is caused by uncertainty, complexity and non-quantifiable factors. However, such methods are variations on stochastic techniques and can never be better than the variation data input. Here these models confront a problem, in that they tend to break the project into stages, being either thirds or more likely 10% increments, and then look at variability at that stage. This approach ignores the relationship between points in a cash flow profile (the sequence) and the variation in 'trend' which may be identified.

Kenley's stochastic approach

The Kenley and Wilson Logit cash flow model, being idiographic, is ideally suited to the development of stochastic models. Stochastic models are those based on probability and therefore prediction using such models provides a central trend and estimate of variation rather than a single answer (deterministic). The stochastic use of the model is shown in Kenley (1999) and Kaka (1999).

The Logit cash flow model

It is interesting to explore the nature of the constants α and β in the Logit model. Khosrowshahi (1991) illustrated the effect of his Q and R constants using the periodic distribution. This showed that for Q (Y_p / C) the height of the periodic distribution, and its rate of increase and decay, varied proportionally. Similarly for R (X_p / n), the position of the peak varied proportionally. Similar results can be seen in the Logit model. Table 3.16 shows an analysis of the α and β from samples S1 and S2 (Kenley and Wilson, 1986). These form the basis for indicating the range of possible α and β values in the following analysis.

Table 3.16 Mean and Standard deviation for α and β from S1 and S2

	Mean	Standard deviation
α	0.2156	0.7365
β	1.6421	0.4732

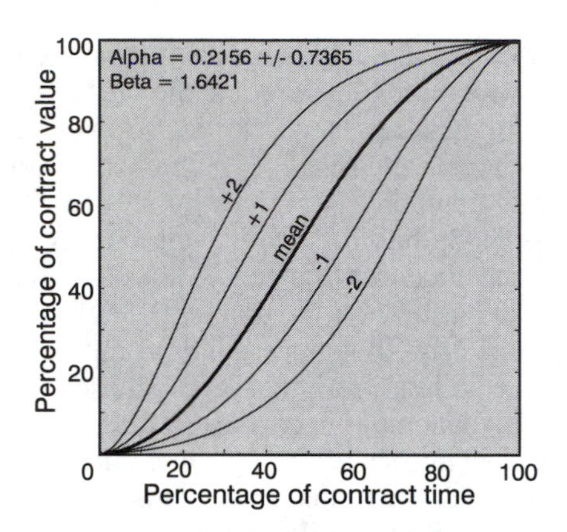

Figure 3.31 The effect of holding β constant and varying α by $\pm 2\sigma$

Figure 3.31 illustrates the effect of the constant α, which can be seen is a position constant—effecting how early the project gets started (with the equivalent delayed completion) but not the maximum rate of expenditure. This is equivalent to Khosrowshahi's *R*.

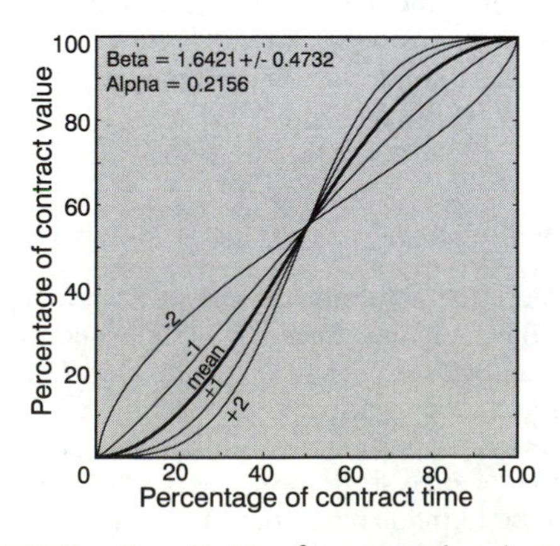

Figure 3.32 The effect of holding β constant and varying α by $\pm 2\sigma$

Figure 3.32 illustrates the effect of the constant β, which it can be seen is a slope or rate constant—affecting how rapidly the project proceeds once started but not the position (or timing). This is equivalent to Khosrowshahi's Q.

It is easy to see from this how an experienced estimator could manually manipulate a cash flow profile by adjusting the values of α and β as suggested by Khosrowshahi (1991). A very wide range of profiles is clearly possible by varying both constants simultaneously.

Comparing the various profiles

An analysis of all the 'standard curves' reveals a very interesting curve envelope. Figure 3.33 shows the standard curves (one of each where there are multiple curves) from the various models analysed. This shows clearly that there is a wide range of standard curves, falling generally between the lower half of the available envelope of Logit curves varying $\alpha \pm 2\sigma$.

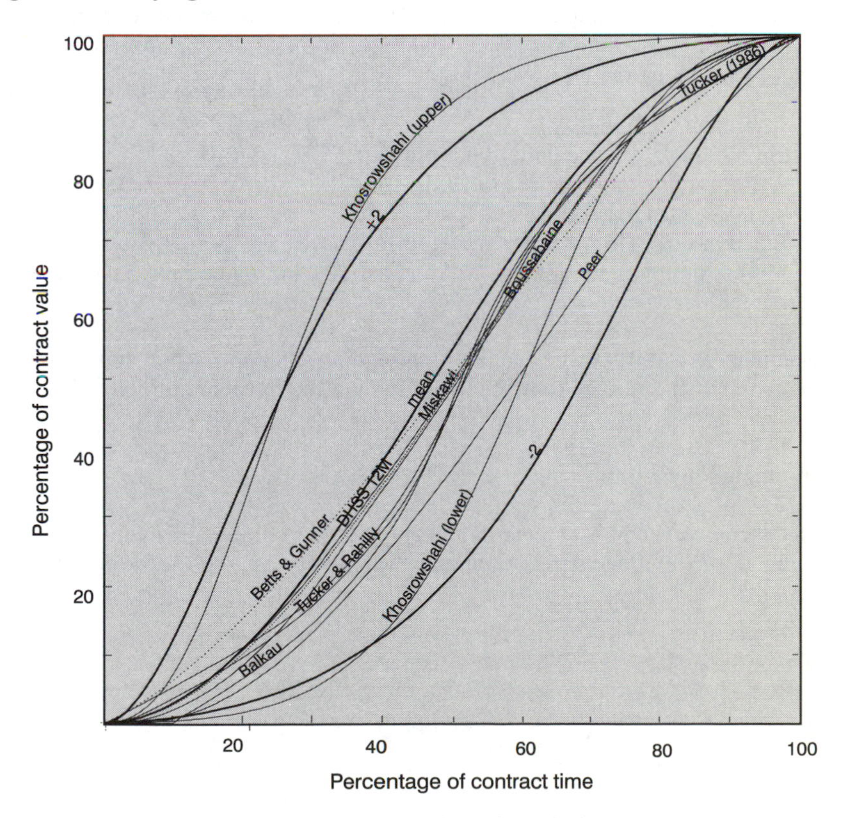

Figure 3.33 The curve envelope of 'standard curves'

This result is to be expected, as standard curves result from aggregating the data of many individual projects, resulting in a fitted curve which describes none of the individual projects upon which it is based. It is possible that many of these standard models are based on projects which have individual profiles quite different from the resultant standard curve, but most likely within the $\alpha \pm 2\sigma$ envelope described here.

These standard curves are sufficiently similar, and all within the lower half of the expected curve envelope, that it becomes necessary to question what is happening. Kosrowshahi is the only researcher expressing curves in the upper bounds. This is probably because of his interest in individual projects rather than standard profiles.

A THEORETICAL COMPARISON OF THE MODELS

Summary

The above models all attempt, by one method or another, to simulate or forecast cash flow profiles for construction projects. Table 3.17 summarises the properties of these models by examining the following:

- Publication
- Control model (forms the underlying profile)
- Number of parameters required to fit the model
- The points of Origin and End
- The Epistemology (Idiographic or Deterministic)
- The Methodology (Deterministic or Stochastic including complex extensions)
- The model identifier.

This summary is a useful way of comparing the different approaches, and follows the approach to the structure of knowledge outlined in the introduction (page 3).

Developments over time

Another useful way to look at these models is charted in Figure 3.34 which shows the development of the models over time and the underlying trend toward idiographic models.

Table 3.17 Properties of cash flow models

Publication	Control model	Number of parameters	Origin and End point properties	Idiographic–nomothetic	Deterministic–stochastic	Comments
Kennedy *et al.* (1970)	Schedule	–	-	Idiographic	Deterministic	Schedule method
Balkau (1975) Bromilow (1978) Bromilow and Davies (1978) Tucker and Rahilly (1982)	Inverse Polynomial by regression	5	Determined by regression	Nomothetic	Deterministic	Bromilow model
Berdicevsky (1978) Peer (1982)	Polynomial by regression	2 to 4	Determined by regression	Nomothetic	Deterministic	Peer model
Betts and Gunner (1993)	Polynomial by regression	3	(0,0) End determined by regression	Nomothetic	Deterministic	Peer model
Kenley and Wilson (1986)	Logit	2	(0,0) and (100,100)	Idiographic	Deterministic and stochastic	Kenley and Wilson
Tucker (1986)	Compound Linear–Weibull	4	(0,0) End determined by iteration	Nomothetic but solved by iteration	Deterministic	Tucker model
Hudson and Maunick (1974) Hudson (1978)	Compound polynomial	2	(0,0) End determined by profile	Idiographic but standards for dollar ranges	Deterministic	DHSS model
Berny and Howes (1982)	Compound polynomial	2	(0,0) End determined by profile	Idiographic but standards for dollar ranges	Deterministic	DHSS re-expressed
Tucker (1988)	Weibull	2	(0,0) End determined by iteration	Nomothetic but solved by iteration	Deterministic	Alternative for DHSS
Miskawi (1989)	Sin	1	(0,0) and (100,100)		Deterministic	
Khosrowshahi (1991)	Exponential plus shaping and distortion	3 +	(0,0) and (100,100)	Idiographic	Stochastic	Kosrowshahi model
Kaka and Price (1993) Evans and Kaka (1998)	Logit	2	(0,0) and (100,100)	Nomothetic, project categories	Deterministic	Kenley and Wilson
Boussabaine and Kaka (1998)	Unknown	Unknown	Unknown	Idiographic	Stochastic—neural networks logic	Boussabaine and Kaka
Boussabaine and Elhag (1999)	Phased polynomial	Fixed	(0,0) and (100,100)	Nomothetic	Stochastic—Fuzzy logic	Boussabaine and Elhag

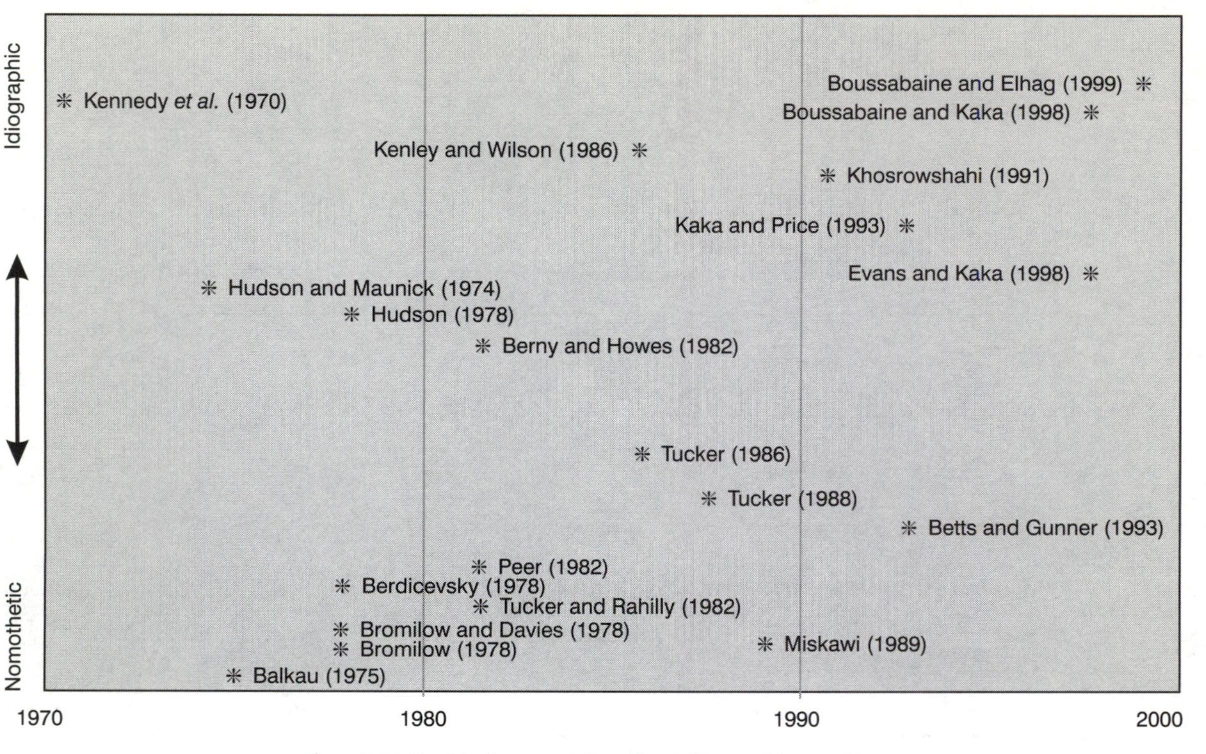

Figure 3.34 Graphical representation of cash flow models over time

Data collection and categorisation

Some of the data collection standards will have varied, for example the difference between focussing on practical completion (Bromilow model) or final payment (most other models). Some of the models may be drawing upon different populations, or there may have been factors at work which correlate with different profiles. Kenley and Wilson (1986) tested the two samples S1 and S2 and did not find they came from significantly different populations. Similar investigations have not yet been carried out on other samples, except Betts and Gunner, who tested data points at given time slices and found that their samples were from single populations. This is not a surprising result because it is an inappropriate statistical test, as the extent of variation about the mean of these discrete points can be expected to be highly variable and does not inform as to the variation of the trend (curve) and whether or not there are different populations in samples of project curves. There have, however, been attempts to identify variables which might influence outcomes.

Hudson provided the first such example, clearly having used project size (dollar value) as a factor. Other researchers to have followed this approach are Singh and Phua (1984—residential, commercial and industrial), Betts and Gunner (1993—commercial, residential, institutional, school), Kaka and Price (1993—contract type: traditional, management, design and build; client: private, public; type: commercial, industrial, buildings, housing, civil; size) and Evans and Kaka (1998—retail).

While it appears valid to undertake a multivariate analysis to look for factors which influence projects, no researchers have yet shown whether or not these results are due to the presence of those factors or rather, due to sampling error. Kaka and Price (1993) argued strongly for focussing on more specific groupings to achieve better fits. Their argument was based on the Kenley and Wilson (1986) version of SDY, which incorrectly measured internal accuracy. Reducing the degrees of freedom to $n-2$ would invalidate their contention for small sample sizes.

Given the wide envelope of curves for individual projects, and the much narrower range of average (standard) curves, it is quite likely that sampling error is sufficient to explain the difference. This contention would explain why researchers have so much difficulty predicting future projects from standard models.

It is important, however, that models of gross cash flow are developed, as they can be used to explore practical project phenomena. As the standard curves appear quite similar, the best model to use is arguably the easiest. For easier calculation and because of its ability, proven in this chapter, to acceptably replicate all other models, the Logit model appears the best. In a way it is a form of magic, each researcher's model has its own approach, is only a model of reality, and is not actually real, but it is the purpose of models to emulate reality. What matters is the ability to adapt the model to different circumstances with greatest ease and reliability.

The remaining chapters of this book will illustrate practical ways in which the Kenley and Wilson Logit model may be used to analyse and explain practical phenomena such as: net cash flows, in-project end date forecasting and . Ultimately, the model can be used to form the basis of an organisational cash flowperformance management model (Kaka 1999) and indeed an industry micro-economic model.

TOOLS FOR CASH FLOW FORECASTING

Using the Logit cash flow model

The Logit cash flow model is designed to be able to fit the cash flow of any project. It will not match the exact data points, but it will approximate the true (smoothed) trend of the cash flow for that project. It is this ability to rapidly and easily calculate individual curves which gives the model its power. In this section, the practical calculation of the α and β from a set of project data will be demonstrated, as well as the calculation of the model's trend line, from which may be calculated the error of the fit (SDY).

Calculating α and β

The following exercises will use the following real project data. This data comes from the construction of a new police station in a country town in Victoria, Australia. Table 3.18 contains the date of payment, the number of calendar days between payments, the periodic payment and the cumulative total at each payment. This is also shown graphically in the Figure 3.35.

Table 3.18 Sample project: data

Date	Calendar days	Periodic value	Cumulative value
10 Aug 97	0		
1 Oct 97	52	$200,000	$200,000
1 Nov 97	31	$400,000	$600,000
1 Dec 97	30	$400,000	$1,000,000
1 Jan 98	31	$300,000	$1,300,000
1 Feb 98	31	$250,000	$1,550,000
1 Mar 98	28	$350,000	$1,900,000
1 Apr 98	31	$350,000	$2,250,000
1 May 98	30	$257,851	$2,507,851

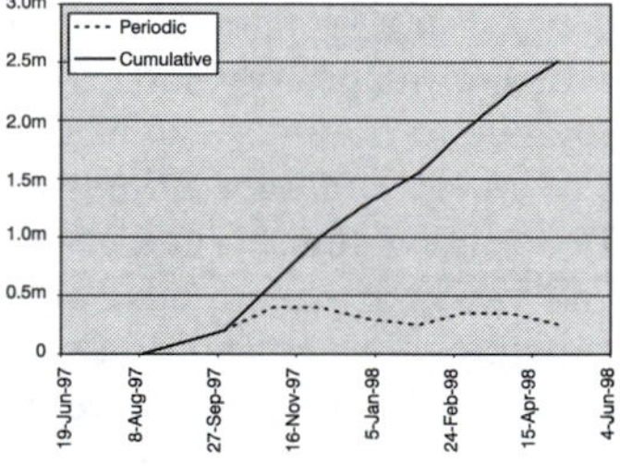

Figure 3.35 Sample curves

The first target is to calculate the α and β constants for a Logit model fit to the data. To achieve this, the first step is to calculate the ratio of the total-time T and total-value V that each progressive payment represents. These may be expressed in percentage terms or ratios. Caution must be exercised here, however, when using spreadsheets to calculate the logit. There is an important difference between the equation for the logit using a ratio of 1.0 and that using 100%, however most spreadsheets when displaying 100% actually store 1.0 in the cell. It is necessary therefore to use the correct ratio version of the formula. For this exercise, the spreadsheets will be storing 1.0 and displaying 100%, therefore the equation where V=1.0 and T= 1.0 will be used (Equations 3.2, 3.4 and 3.7).

$$\ln\left(\frac{v}{1-v}\right)=\alpha+\beta\left(\frac{t}{1-t}\right) \qquad \text{(3.2 repeated)}$$

$$\text{or } v=\frac{F}{1+F} \text{ where } F=e^{\alpha}\left(\frac{t}{1-t}\right)^{\beta} \qquad \text{(3.4 repeated)}$$

$$Y=\alpha+\beta X \text{ where } Y=\ln\left(\frac{v}{100-v}\right) \text{and } X=\ln\left(\frac{t}{100-t}\right) \qquad \text{(3.7 repeated)}$$

From the ratio or percentage data, the next step is to calculate the logit of all data points, both *X* and *Y*. The *Logit* is given by the general form in Equation 3.1 (note that the function Ln() is the natural logarithm, or $\mathrm{Log}_e()$.

$$Logit=\ln\left(\frac{z}{1-z}\right) \qquad \text{(3.1 repeated)}$$

The two columns of logit data are the transformed data required for the linear regression.

Table 3.19 Sample project: percentage calculation

Calendar days	Cumulative days	Periodic value	Cumulative value	*t*	*v*
0	0			0.00%	0.00%
52	52	$200,000	$200,000	19.70%	7.97%
31	83	$400,000	$600,000	31.44%	23.92%
30	113	$400,000	$1,000,000	42.80%	39.87%
31	144	$300,000	$1,300,000	54.55%	51.84%
31	175	$250,000	$1,5500,000	66.29%	61.81%
28	203	$350,000	$1,900,000	76.89%	75.76%
31	234	$350,000	$2,250,000	88.64%	89.72%
30	264	$257,851	$2,507,851	100.00%	100.00%

Note: $T = 264$, $V = \$257, 851$

Table 3.19 includes the periodic data, the cumulative data and the percentages *t* and *v*. These may then be used to substitute *t* and *v* for *z* in Equation 3.1. The logits are calculated in Table 3.20.

The logit transformations are calculated from the percentage data, but cannot be calculated for 0% or 100%. These logits are therefore ignored. This was a major concern to Skitmore (1992) who altered the maximum values to 100.1%, however, the model assumes these points, as they are an inherent property of the Logit model, and therefore no such adjustment is required and those points may be safely ignored.

Table 3.20 Sample project: logit calculation

t	v	X logit (v)	Y logit (t)	
0.00%	0.00%			
19.70%	7.97%	–1.4053	–2.4458	
31.44%	23.92%	–0.7797	–1.1568	
42.80%	39.87%	–0.2899	–0.4107	
54.55%	51.84%	0.1823	0.0735	See Table 3.21
66.29%	61.81%	0.6762	0.4813	
76.89%	75.76%	1.20233	1.1397	
88.64%	89.72%	2.05412	2.1663	
100.00%	100.00%			

It was established earlier that a better fit is usually achieved by ignoring data points outside the 10% exclusion range. The shaded cells lie outside these ranges and would normally be deleted from the analysis. For this exercise they will be left in, and then later the SDY of the result with them left out can be compared.

Table 3.21 Sample project: manual logit calculation

X logit (v)	Y logit (t)	$X-\bar{X}$	$Y-\bar{Y}$	$(X-\bar{X})(Y-\bar{Y})$	$(X-\bar{X})^2$
–1.4053	–2.4458	–1.6396	–2.4240	3.9744	2.6884
–0.7797	–1.1568	–1.0139	–1.1350	1.1509	1.0281
–0.2899	–0.4107	–0.5242	–0.3889	0.2039	0.2748
0.1823	0.0735	–0.0520	0.0953	–0.0050	0.0027
0.6762	0.4813	0.4419	0.5031	0.2223	0.1952
1.2023	1.1397	0.9680	1.1615	1.1243	0.9371
2.0541	2.1663	1.8198	2.1881	3.9819	3.3118
0.2343	-0.0218	<=Mean	Ave=>	10.6528	8.4381

Using a spreadsheet, it is a simple calculation to derive α and β from the logit transformations. The spreadsheet functions are as follows:

$$\alpha = INTERCEPT\left(Y_1 : Y_n, X_1 : X_n\right) \tag{3.52}$$

$$\alpha = SLOPE\left(Y_1 : Y_n, X_1 : X_n\right) \tag{3.53}$$

which will yield $\alpha = -0.3176$ and $\beta = 1.2625$. It is sometimes, however, useful to calculate α and β manually. The equations are:

$$\beta = \frac{\sum\left[(X-\overline{X})(Y-\overline{Y})\right]}{\sum(X-\overline{X})^2} \qquad \text{(3.8 repeated)}$$

$$\text{and } \alpha = \overline{Y} - \beta\overline{X} \qquad \text{(3.9 repeated)}$$

This is best illustrated using Table 3.21, where Equations 3.8 and 3.9 will also yield $\alpha = -0.3176$ and $\beta = 1.2625$.

Testing the fitted model

Testing the fitted model against the data requires calculating for each data point (t, v), the modelled point (t, v') where v' is the calculated value at time t. This is calculated using Equation 3.4 repeated above. This is easiest in a spreadsheet using a function in each cell.[4] The function developed is named here CumulativeCF(), but could be given any non-reserved name. The code in Table 3.22 is in *Visual Basic for Applications*, but could easily be adapted to other programming languages.

Table 3.22 Code for the function CumulativeCF()

```
Function CumulativeCF(TotalValue, TotalTime, StartDate, Alpha, Beta, ThisDate)
  Days = ThisDate - StartDate
  If Days <= 0 Then
    CumulativeCF = 0
  ElseIf Days < TotalTime Then
    TimeP = Days / TotalTime
    F = Exp(Alpha) * (TimeP / (1 - TimeP)) ^ Beta
    ValueP = F / (1 + F)
    CumulativeCF = ValueP * TotalValue
  Else
    CumulativeCF = TotalValue
  End If
End Function
```

The code in CumulativeCF() is generic and allows either introduction of real data, percentage data or ratio (1.0) data. If using ratio data for the sample above (and therefore to output ratio data (usually formatted as percentages) then the function would be called as follows:

$$v' = CumulativeCF\left(1, 1, 0, 0.3176, 1.2625, t\right) \qquad (3.54)$$

Simpler versions of the function could be developed where a generic version is not required.

[4] This function is worth developing, for example in Excel using Visual Basic, as it is able to be used repetitively for many processes when working with cash flows.

Calculation of the goodness-of-fit of the model uses the SDY equation, with $(n-2)$ degrees of freedom. SDY should be calculated on percentage data to ensure comparability. The SDY Equation is 3.13, and Table 3.23 shows the fitted value v' and the calculation of SDY.

$$SDY = \sqrt{\frac{\sum (Y - Y_E)^2}{n-2}} \qquad \text{(3.13 repeated)}$$

Table 3.23 Sample project: original data, fitted values and calculation of SDY

t	v	v'	$v-v'$	$(v-v')^2$
0%	0%			
19.70%	7.97%	10.99%	–3.02%	0.09%
31.44%	23.92%	21.39%	2.54%	0.06%
42.80%	39.87%	33.55%	6.33%	0.40%
54.55%	51.84%	47.82%	4.02%	0.16%
66.29%	61.81%	63.09%	–1.28%	0.02%
76.89%	75.76%	76.86%	–1.10%	0.01%
88.64%	89.72%	90.68%	–0.97%	0.01%
100.00%	100.00%	100.00%		
				SDY = 3.28%

That this is not a great fit for the data is clear from the value of SDY. A visual inspection of the data (Figure 3.36) shows why this is so. First, there are not many data points, resulting in only five degrees of freedom. Secondly, the project did not progress very smoothly. This is common on small projects.

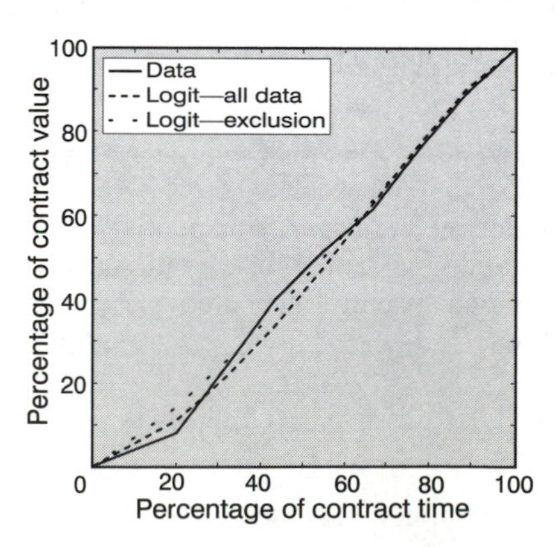

Figure 3.36 Sample project: forecast data and fitted curves, SDY = 3.28%, 2.85%

Removing the data outside the exclusion range (the shaded data in Tables 3.20 and 3.21) results in an improvement of SDY from 3.28% to 2.85%. The resultant curve, shown as 'Logit—exclusion' on Figure 3.36, ignores the first data point and improves the fit for the remaining points.

This can be extended further, as an exercise, using the techniques above. It can now be revealed that the above sample project was in fact the forecast cash flow curve provided by the builder to the client at the start of the project. This may now be compared with the actual cash flows for the same project. Table 3.24 provides the actual data and the results.

Table 3.24 Sample project: actual data and results

Date	Calendar days	Periodic value	Cumulative value	t	v	v'
10 Aug 97	0			0%	0%	0%
18 Sep 97	39	$209,655	$209,655	13.73%	8.03%	7.94%
16 Oct 97	67	$248,940	$458,595	23.59%	17.57%	17.35%
30 Oct 97	81	$156,248	$614,843	28.52%	23.56%	22.84%
13 Nov 97	95	$129,865	$744,708	33.45%	28.54%	28.76%
27 Nov 97	109	$361,685	$1,106,393	38.38%	42.40%	34.99%
11 Dec 97	123	$185,878	$1,292,271	43.31%	49.52%	41.45%
18 Jan 98	161	$204,202	$1,496,473	56.69%	57.35%	59.32%
29 Jan 98	172	$143,562	$1,640,035	60.56%	62.85%	64.37%
12 Feb 98	186	$140,762	$1,780,797	65.49%	68.24%	70.60%
26 Feb 98	200	$223,858	$2,004,655	70.42%	76.82%	76.50%
12 Mar 98	214	$120,253	$2,124,908	75.35%	81.43%	81.99%
26 Mar 98	228	$171,460	$2,296,368	80.28%	88.00%	86.99%
16 Apr 98	249	$209,358	$2,505,726	87.68%	96.02%	93.40%
7 May 98	270	$44,833	$2,550,559	95.07%	97.74%	98.18%
2 Jun 98	284	$58,959	$2,609,518	100%	100%	100%

Note: SDY for all data = 3.17, SDY with data excluded = 2.88

Excluding the data outside the exclusion ranges (shaded cells in Table 3.24) results in $\alpha=0.0562$ and $\beta=1.2957$ with $SDY=2.88\%$ (using all the data results in $\alpha=0.0159$ and $\beta=1.3421$ with $SDY=3.17\%$). The actual data with resultant logit is charted in Figure 3.37, together with the forecasts.

In the Chapter 4, 'Earned Value', the flexibility of the model will be highlighted by taking this 'before' and 'after' analysis to forecast the end date progressively through the project, using the Kenley and Wilson Logit model.

In this instance the forecast and the actual cash flows are similar despite the very different timing of payments. Other projects have different end-dates, different values, or different parameters for the curves. These issues become critical when assessing project performance.

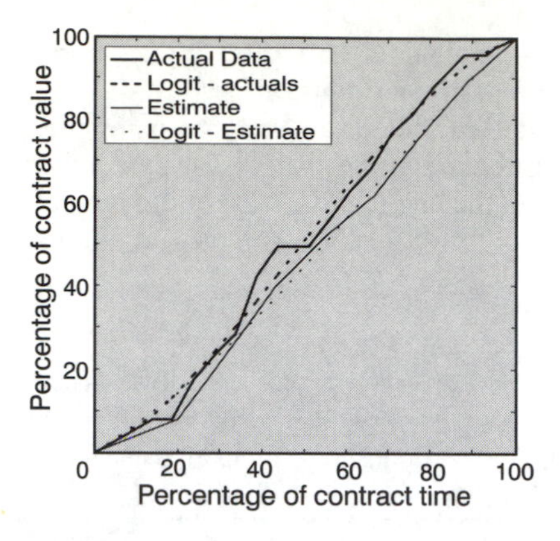

Figure 3.37 Sample project: actual data and fitted curves, SDY = 3.17%, 2.88%

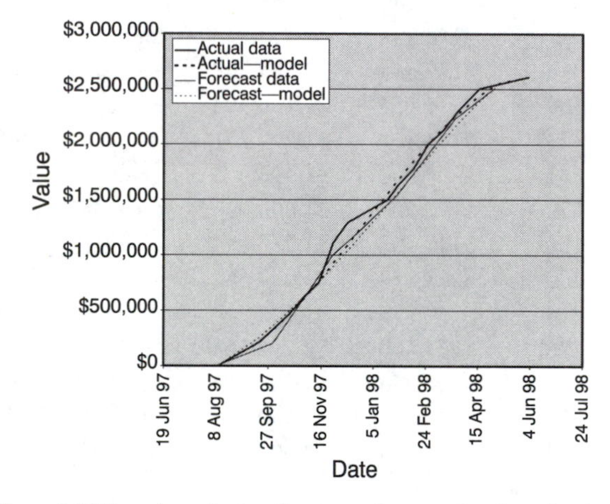

Figure 3.38 Sample project: estimates and source data in real values

CONCLUSION

The last thirty or more years has seen the growth of a plethora of models to describe the progressive relationship between time and cost on construction projects. Such is the extent of variation in this relationship, however, that these models have resorted to attempting to find a model which minimises variation about the model.

Those models which have tried to identify standard curves have largely been frustrated by the seemingly huge range of factors which effect the outcome, most essentially unpredictable. There have been two fundamental approaches to the problem, those based on the concept of a standard or average curve and those which accept the individual variation as a property of the relationship. Recent methods have recognised this characteristic of project cash flows and have attempted to manage the variation through the application of stochastic and fuzzy approaches based on complexity.

It is an exciting time in this field, particularly as researchers come to grapple with the even more complex issues of net cash flow (Chapter 6) and organisational cash flow (Chapter 7). Of immediate value has been the new emphasis on performance measurement, which uses the cash flow methods and applies them to monitoring the in-project performance (Chapter 4).

REFERENCES

Ashley, D. B. and Teicholz, P. M. (1977). 'Pre-estimate cash flow analysis'. *Journal of the Construction Division, ASCE*, Proc. Paper 13213, **103** (CO3): 369–379. See also discussion by Gates, M. and Scarpa, A. (1978), **104** (CO1): 111–113, and closure by Ashley *et al.* (1978), **104** (CO4): 554.

Ashton, W. D. (1972). *The Logit Transformation*. Griffins statistical monographs and courses, No. 32. Griffin, London.

Balkau, B. J. (1975). 'A financial model for public works programmes'. Paper to National ASOR Conference, Sydney, August 25–27.

Berdicevsky, S. (1978). Erection cost flow analysis. Unpublished MSc Thesis, Technion, Israel Institute of Technology, Haifa, Israel.

Berny, J. and Howes, R. (1982). 'Project management control using real time budgeting and forecasting models'. *Construction Papers* **2**: 19–40.

Betts, M. and Gunner, J. (1993). *Financial Management of Construction Projects: Cases and Theory in the Pacific Rim*. Longman, Singapore, Ch. 11: 207–227.

Boussabaine, A. H. and Elhag, T. (1999). 'Applying fuzzy techniques to cash flow analysis'. *Construction Management and Economics* **17**: 745–755.

Boussabaine, A. H. and Kaka, A. P. (1998). 'A neural networks approach for cost-flow forecasting'. *Construction Management and Economics* **16**(4): 471–479.

Boussabaine, A. H., Thomas, R. and Elhag, M. S. (1999). 'Modelling cost-flow forecasting for water pipeline projects using neural networks'. *Engineering, Construction and Architectural Management* **6**(3): 213–224.

Bromilow, F. J. (1978). 'Multi-project planning and control in construction authorities'. *Building Economist*, March, 208–213, also September, 67–70.

Bromilow, F. J. and Davies, V. F. (1978). 'Financial planning and control of large programmes of public works'. Second International Symposium on Organisation and Management of Construction, *Organising and Managing Construction*. Technion, Israel Institute of Technology, Haifa, Israel.

Bromilow, F. J. and Henderson, J. A. (1977). *Procedures for Reckoning the Performance of Building Contracts*, 2nd edn. CSIRO, Division of Building Research, Highett, Australia.

Cooke, B. and Jepson, W. B. (1979). *Cost and Financial Control for Construction Firms*. London, Macmillan.

Davis, R. (1991). *Construction Insolvency*. London, Chancery Law Publishing.

De Groot, A. D. (1969). *Methodology - Foundation of Inference and Research in the Behavioral Sciences*. Mouton & Co., Belgium.

Drake, B. E. (1978). 'A mathematical model for expenditure forecasting post contract'. In Proceedings of the Second International Symposium on Organisation and Management of Construction, *Organising and Managing Construction*. Technion, Israel Institute of Technology, Haifa, Israel, **2**: 163–83.

Evans, R. C. and Kaka, A. P. (1998). 'Analysis of the accuracy of standard/average value curves using food retail building projects as case studies'. *Engineering, Construction and Architectural Management* **5**(1): 58–67.

Fleming, Q. W. and Fleming, Q. J. (1991). *A Probus Guide to Subcontract Management and Control.* Probus publishing company, Chicago.

Hardy, J. V. (1970). Cash flow forecasting for the construction industry. MSc Report; Dept. of Civil Engineering, Loughborough University of Technology, UK.

Hoaglin, D. C., Mosteller, F. and Tukey, J. W. (1983). *Understanding Robust and Exploratory Analysis*. John Wiley & Sons, Inc., New York.

Hudson, K. W. (1978). 'DHSS expenditure forecasting method'. *Chartered Surveyor—Building and Quantity Surveying Quarterly* **5**: 42–45.

Hudson, K. W. and Maunick, J. (1974). Capital expenditure forecasting on health building schemes, or a proposed method of expenditure forecast. Research report, Surveying Division, Research Section, Department of Health and Social Security, UK.

Ireland, V. B. E. (1983). The role of managerial actions in the cost, time and quality performance of high rise commercial building projects. Unpublished PhD Thesis, Sydney, Faculty of Architecture, University of Sydney.

Jepson, W. B. (1969). 'Financial control of construction and reducing the element of risk'. *Contract Journal*, April 24, 862–864.

Kaka, A. P. (1999). 'The development of a benchmark model that uses historical data for monitoring the progress of current construction projects'. *Engineering, Construction and Architectural Management* **6**(3): 256–266.

Kaka, A. P. and Price, A. D. F. (1993). 'Modelling standard cost commitment curves for contractors'. *Construction Management and Economics* **11**: 271–283.

Kaka, A. P. and Price, A. D. F. (1994). 'A survey of constractor's corporate planning and financial budgeting'. *Building Research and Information* **22**(3): 174–182.

Kenley, R. (1999). 'Cash farming in building and construction: a stochastic analysis'. *Construction Management and Economics* **17**(3): 393–401.

Kenley, R. and Wilson, O. D. (1986). 'A Construction Project Cash Flow model—an Idiographic Approach'. *Construction Management and Economics* **4**: 213–232.

Kennedy, W. B., Anson, M., Myers, K. A. and Clears, M. (1970). 'Client time and cost control with network analysis'. *The Building Economist* **9**: 82–92.

Kerr, D. (1973). Cash flow forecasting. MSc report; Department of Civil Engineering, Loughborough University of Technology, UK.

Khosrowshahi, F. (1991). 'Simulation of expenditure patterns of construction projects'. *Construction Management and Economics* **9**(2): 113–132.

McCaffer, R. (1979). 'Cash flow forecasting'. *Quantity Surveying*, August, 22–26.

Miskawi, Z. (1989). 'An S-curve equation for project control'. *Construction Management and Economics* **7**: 115–124.

Nazem, S. M. (1968). 'Planning contractor's capital'. *Building Technology and Management* **6**, 256–260.

Padman, R and Smith-Daniels, D. E. (1993). 'Early-tardy cost trade-offs in resource constrained projects with cash flows: An optimisation-guided heuristic approach'. *European Journal of Operational Research* **64**: 295–311.

Peer, S. (1982). 'Application of cost-flow forecasting models'. *Journal of the Construction Division*, ASCE, Proc. Paper 17128, **108** (CO2): 226–32.

Peterman, G. G. (1973). 'A way to forecast cash flow'. *World Construction*, October: 17–22.

Pilcher, R. (1966). Budgetary and cost control for civil engineering construction allied to modern methods of programming. MSc Thesis, Loughborough University of Technology, Loughborough.

Reinschmidt, K. F. and Frank, W. E. (1976). 'Construction cash flow management system'. *Journal of the Construction Division*, ASCE, Proc. Paper 12610, **102** (CO4): 615–627.

Rescher, N. (1985). *The Strife of Systems*. Pittsburgh, P.A., University of Pittsburgh Press.

Runyan, W. M. (1983). 'Idiographic goals and methods in the study of lives'. *Journal of Personality* **51**: 413–437.

Sidwell, A. C. and Rumball, M. A. (1982). 'The prediction of expenditure profiles for building projects'. In *Building Cost Techniques: New Directions*. P. S. Brandon Ed., London, E. & F.N. Spon.

Singh, S. and Phua, W. W. (1984). 'Cash flow trends for high rise building projects'. In Proceedings of the 4th International Symposium on Organisation and Management of Construction, *Organising and Managing Construction*, University of Waterloo, Canada.

Skitmore, M. (1992). 'Parameter prediction for cash flow forecasting models'. *Construction Management and Economics* **10**: 397–413.

Spirer, H. F. (1975). *Business Statistics: a Problem Solving Approach*. Illinois, Richard D. Irwin.

Tucker, S. N. (1986). 'Formulating construction cash flow curves using a reliability theory analogy'. *Construction Management and Economics* **4**(3): 179–188.

Tucker, S. N. (1988). 'A single alternative formula for Department of Health and Social Security S-Curves'. *Construction Management and Economics* **6**(1): 13–23.

Tucker, S. N. and Rahilly, M. (1982). 'A single project cash flow model for a microcomputer'. *Building Economist*, December: 109–15.

Tucker, S. N. and Rahilly, M. (1985). FINCASH Computer user manual. Division of Building Research, Commonwealth Scientific and Industrial Research Organisation.

Velleman, P. F. and Hoaglin, D. C. (1981). *Applications, Basics and Computing of' Exploratory Data Analysi*s. Boston, Duxbury Press.

Weisberg, S. (1980). *Applied Linear Regression*. New York, John Wiley & Sons.

Wonnacott, T. H. and Wonnacott, R. J. (1972). *Introductory Statistics for Business end Economics*. New York, John Wiley & Sons.

Zoisner, J. (1974). Erection cost flow analysis in housing projects as a function of its size and construction time. MSc Thesis, Technion—Israel Institute of Technology, Haifa, Israel.

4

Managing through earned value

INTRODUCTION

Researchers have repeatedly given as a major reason for investigating construction project gross cash flows, 'that such knowledge can help prevent financial failure'. Although such knowledge is, on its own, unlikely to prevent failure, there are many benefits which accrue from monitoring and managing a contractor's cash flow. These benefits can accrue to either the contractor or the client. It might be more reasonable to argue that knowledge of gross cash flow may lead to prevention of project failure. Certainly, it may be of considerable assistance to alerting management to problems on projects.

Kaka (1999) argued that 'one of the key objectives of managers in construction firms, is to control cost and ensure projects will be completed within contracted duration'. There are a plethora of techniques for doing so, usually involving some form of direct monitoring and reporting. For example, monitoring costs against budgets, with 'expected to complete' totals being calculated progressively, is a direct form of monitoring project cost. Similarly, maintaining a detailed and accurate project activity schedule, with continuous progress updating, is a direct form of monitoring project time. It is less well known that integrated tools may be used to achieve these objectives, as well as indirect monitoring. 'Earned value' describes a suite of related techniques which may be used to achieve either integrated or remote monitoring of cost and time or both. This term has been appropriated from 'earned value performance management' discussed later in this chapter, because it remains the single best term to describe the use of progress cash flow to measure project performance.

Integrative techniques are sometimes referred to as the measurement of performance of the whole project. This is a slight exaggeration, but it does express the importance of integrated techniques. An earned value analysis is able to monitor the complex interaction between cost and time in order to provide feedback to management. Done in a timely manner through the use of a control cycle, this has the capacity to greatly improve the management of projects.

Remote monitoring techniques are powerful through their ability to allow management intervention or interaction by those not directly involved in the management of a project. The most frequently used example of this is the client, who can use remote monitoring techniques to assess a contractor's performance. A less well known example, but arguably equally important (if sometimes less popular on site) is the central financial management of the contractor. Not many contractor's project managers appreciate the active intervention of head office in the management of projects, therefore the ability to monitor and report performance remotely is a powerful cross-check for management. This is even more so if the process is automated and built into accounting and other reporting systems.

Earned value techniques monitor the value of work performed, the cost of that work against budget and the time taken to produce that work against schedule. This integrated approach greatly increases the value of information derived from sources which, managed separately, do not reveal the true performance of a project.

INTEGRATED AND REMOTE PERFORMANCE MANAGEMENT CONCEPTS

Controlling through performance management

Cybernetics and systems

The management of projects through monitoring and control is a control–feedback system which owes much to Wiener's concept of 'cybernetics'. In 1948, Norbert Wiener published his *Cybernetics: Or Control and Communication in the Animal and the Machine* (Wiener, 1996). In this, he outlined the historical development of a new school of thought that encompassed electronics and electrical engineering, mechanical engineering, mathematics, biology and sociology. This was cybernetics, covering '...an essential unity of the set of problems centering about communication, control, and statistical mechanics, whether in the machine or in living tissue'.

Cybernetics comes from the Greek for steersman. The earliest form that may be identified is that of the governor in mechanical systems, described in 1868. Wiener used the analogy of the steering engines of a ship, as an early and well-developed form of feedback mechanism.

Cybernetics has developed into a rigorous mathematical and scientific body of knowledge and we have benefited greatly from it. It has provided the theory for much of what we take for granted, such as radio (noise removal), thermostats, automatic piloting systems, self-propelled missiles, ultra-rapid computing machines, prosthetics, bionic ears, and anti-aircraft guns.

Automata that replicate the biological world

The extraction of Dr Who's Cyberman, and the common picture of robotics, relates to the realisation that cybernetics applies to automata that replicate the biological world. Machines may be constructed with sensors through which impressions are received (replicating ears, eyes, etc.), locators which record the position in space and time of the impression (corresponding to kinesthetic organs and proprioceptors in the human system), processors to analyse the inputs (brain) and effectors such as solenoids, motors, heaters (action in the animal system—limb movement etc.). The impressions may be stored for later action (analogous to memory) and '...as long as the automation is running, its very rules of operation are susceptible to some change on the basis of the data which have passed its receptors in the past, and this is not unlike the process of learning' (Wiener, 1996).

Wiener left to others the extraction from his scientific rigour to less precise organisational theory. As he said 'It is certainly true that the social system is an organisation like the individual, that it is bound together by a system of communication, and that it has dynamics in which circular processes of a feedback nature play an important part'.

Control systems for construction management

The circular feedback processes which the concept of cybernetics entails is fundamental to the successful management of construction projects. We do not normally use such terminology, preferring rather the use of terms such as 'control loop', 'monitoring cycle', or even 'Single Loop Learning' (Stacey, 1996), although Betts and Gunner (1993) differentiate between homoeostatic systems with feedback loops and cybernetic systems with adaptive standards 'where corrective action can influence both the input to a system and the process within the system itself'. Whatever the terminology or the model, the process is one of setting a plan, performance, monitoring performance against the plan, initiating corrective action and repeating the cycle. This simple control cycle is illustrated in Figure 4.1.

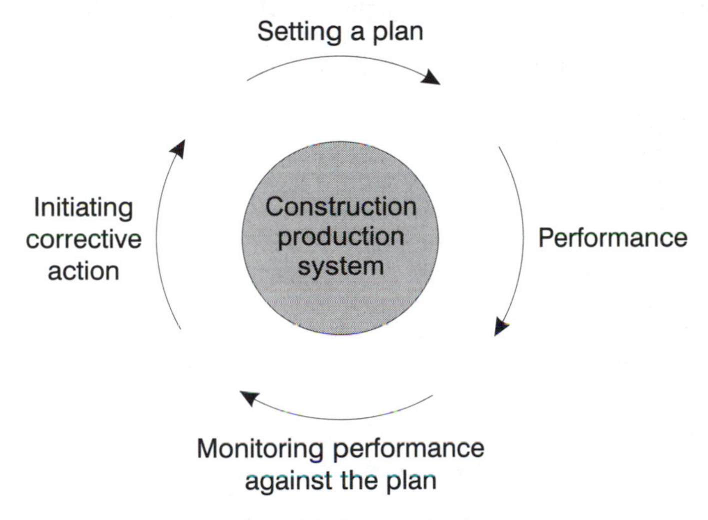

Figure 4.1 The control cycle

The important thing to remember with the control cycle, is that it only functions in a supportive environment. When management does not initiate or maintain the cycle, feedback fails. In these circumstances, all too common with regard to project performance management, performance is only as good as the first or latest plan.

The 'steersman' and the control–cost tradeoff

It is useful to go back to the roots of cybernetics to understand the real value of control systems. This helps to understand a simple concept that can become confusing to the point of distraction. In turn it helps to understand the practical cost implications of control systems—a very real barrier to their implementation or success.

Imagine you are driving a car. You are on a long, remote, generally straight, tree-lined road, at night. Your alternator is broken and your battery is almost flat. If you turn the lights off it is pitch black; with them on you can see well out in front. You have to reach a destination, in one piece, on time. This is a classic construction problem. You have a budget (limited battery life) a quality target (not damaging your vehicle), a budget (the road) and a target (the destination on time).

You cannot leave the lights on because the battery will rapidly go flat, leaving you stranded. You cannot leave them off because you can't see the road. The trade-off is to turn them on with sufficient frequency to be able to: (a) see where you are in relation to the road, (b) correct the steering and avoid danger, and (c) plan the path further down the road sufficiently to survive until the lights are next turned on again for a further look. Too many looks and you run out of budget (power). Too few looks and quality suffers (damage).

This is the control–cost trade-off. Another good demonstration of this tradeoff is to mark out a path between obstacles, such as furniture, and then visualise that path to be walked blindfolded. How many times do you need to peek to minimise the number of bumps into obstacles. There is an inverse relationship between the number of peeks and the number of bumps. All that needs now to be worked out is the relative value of a peek and a bump to solve for the optimum number of looks to minimise cost.

There is a cost to monitoring (looking). There is a cost to the consequence of not monitoring (bumps or damage). If the cost of monitoring is too high, then there is a risk that it will not be undertaken. Similarly, if it is believed that the consequences of not monitoring are low, then monitoring may be ignored. For example, when driving a car, looks are very cheap and consequences of bumps are very expensive. Thus drivers constantly monitor the road in real time. In situations where an inspection is relatively expensive, looks may be rare and may be delayed (that is, not in real time—for example through some form of periodic reporting system, possibly monthly). The control–cost tradeoff may be represented diagrammatically as in Figure 4.2.

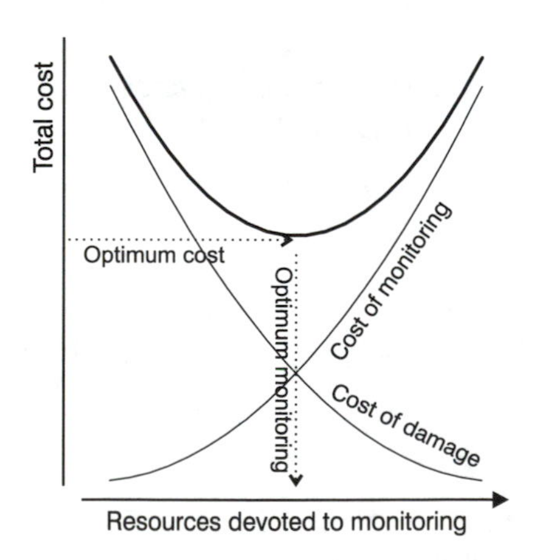

Figure 4.2 The control–cost trade off

The cost of monitoring is, for the financial management and progress of construction projects, the cost of implementing and resourcing the monitoring systems. For example, it is possible to employ staff to track every order, every movement on site, the progress of all events, and to continually update schedules and cost–control systems. This would normally be prohibitively expensive.

The cost of not monitoring is, for the financial management and progress of construction projects, the cost of delayed projects or cost over-runs. This cost may be significant through missed deadlines or cost targets.

The performance management concepts being discussed in this chapter are generally neither simple nor cheap. This discourages their implementation. In the first instance, remote management methods have been complex, poorly researched and expensive to implement. Direct integrated methods have been technically complex and labour intensive to manage.

The performance management problem

A project falls behind?

Imagine a very frustrated project administrator, who has set out a cash flow forecast for a project, using the DHSS standard cash flow model derived from the DHSS standard curve for £12 million, which had been considered appropriate for a $14 million project. The cash flow profile was provided to the client and head office management. For the first two months of the project all seemed fine, and then things had gone horribly wrong, despite the best efforts of the site team. It was a difficult project, but not abnormally so, and the site team was just having difficulty spending the money fast enough. At only 40% of the way through the project, the cash flow profile was already behind the forecast by some 20%. With the value of work in progress, and thus the claims from the client, below the forecast, this had upset the site manager, the project manager, the company accountant, and indeed the client.

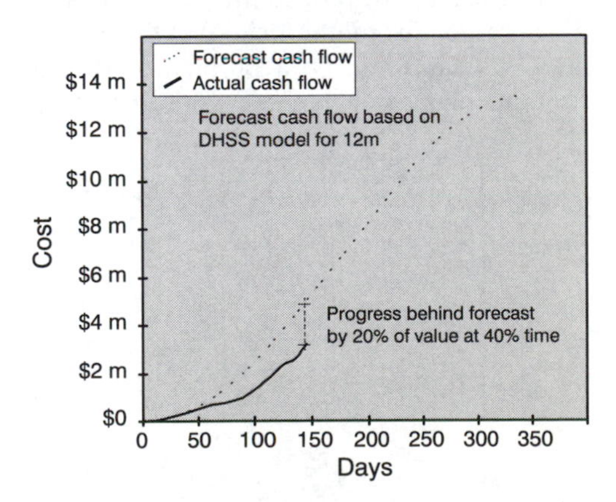

Figure 4.3 Actual cash flow lagging behind forecast (adapted from Kenley, 2001)

This situation, based on a real project, is illustrated in Figure 4.3. Project progress is clearly well behind the forecast in some way.

Interpreting variation from the forecast

Initial reactions were extremely negative, as it was perceived that the project had fallen behind schedule, indicating that the project was going to finish late. The project administrator, however, felt that this was unfair, and that this interpretation was open to alternative analysis. First, it relied on the assumption that the original cash flow forecast was indeed correct. Secondly, it relied on the assumption that the project duration was correctly forecast. And finally, it relied on the assumption that the project cost had been correctly estimated. At this stage it is very difficult to see whether any, or even a combination, of these factors was responsible for the problem. The options may be summarised as:

1. Project is on schedule, but cash flow profile was incorrectly estimated.
2. Project end-date was in error, more time required.
3. Project behind schedule—time over-run likely without remedial action.
4. Project is below budget—total project cost likely to reduce.
5. Project cost estimate was wrong, it was overestimated.
6. A combination of these.

1. Cash flow profile was incorrectly estimated

Figure 4.4 shows the first alternative outcome, in which the problem arises from the incorrect estimation of the cash flow profile, rather than the progress of the work or incorrect evaluation of time or cost. In this scenario, there is nothing wrong with the management of the project. Rather, the problem lies with the estimation of the cash flow profile when it was originally established. Practically, this means that the early effort delivered little value. On a cost profile this would mean that it was relatively inexpensive. On a value profile, this might mean that the early work was not valued highly in the contract. This can often be a problem with projects which require detailed preparation without placing chargeable work in place.

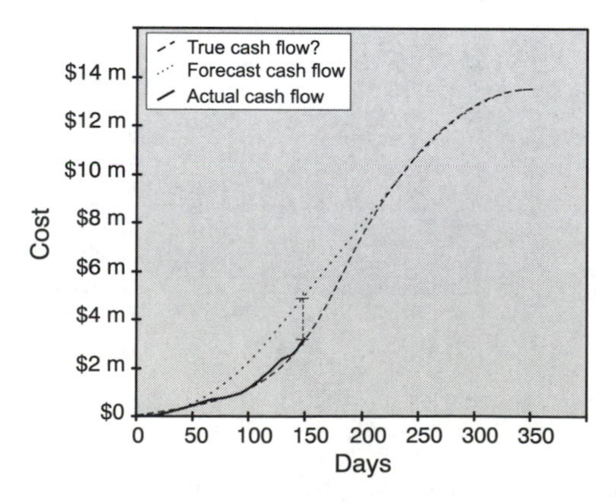

Figure 4.4 Correcting by changing the forecast cash flow profile (adapted from Kenley, 2001)

Given that the cash flow profile was derived from a standard curve, this is a likely outcome, as it has been shown that there is considerable variation in cash flow profiles between individual curves. If this is the case, then no remedial action is required. In fact, spending money to accelerate the works to achieve the original profile might risk increasing the total cost of the project.

Managers who do not take remedial action on such a project are, by default, assuming this to be the correct, albeit risky, interpretation.

2. Project end-date was forecast in error

The second alternative, that the project time estimate is wrong, is illustrated in Figure 4.5. This must be a frequent occurrence given the extraordinary frequency of time increases on projects. In fact the Bromilow cash flow model assumed that the average profile would complete at 117% of practical completion. Bromilow and Henderson (1977) identified that projects usually over-run their end dates (refer to the discussion of time performance ratios in Chapter 5).

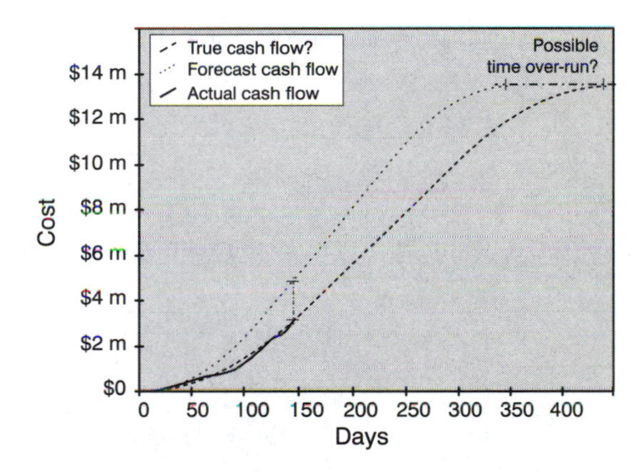

Figure 4.5 Correcting by changing the forecast end-date (adapted from Kenley, 2001)

There is a fine line between this explanation and the third, that the project is behind schedule, and it may in fact be that the correct answer belongs to the responsible individual with the loudest voice and the most persistence. Whichever argument prevails, the diagram remains the same for both.

3. Project behind schedule—remedial action required

The most common belief is that remedial action (accelerating the works) will avoid the end result (time over-run) given in Figure 4.5. The fear that the project duration will blow-out is just cause for remedial action. Such action might involve revising the project schedule by providing additional resources to increase the rate of progress and thus restore the project to schedule. This outcome is outlined in Figure 4.6.

There is likely to be a cost associated with accelerating the works through lost productivity. Horner and Duff (2001) argue this loss of productivity could come from overtime, increased gang sizes, unplanned work or increased waste in labour and materials. Management must make an assessment of the relative costs and benefits of accelerating the work.

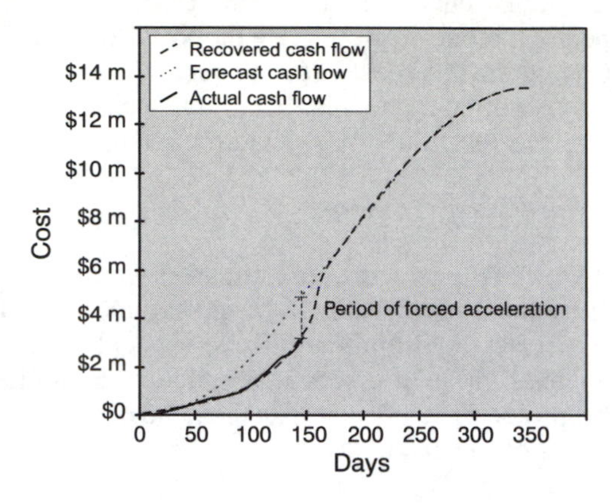

Figure 4.6 Correcting by accelerating progress (adapted from Kenley, 2001)

4. Total project cost likely to reduce

The fourth alternative, that the cost estimate was correct but costs are being saved through good letting of work packages and perhaps changed construction methods, is illustrated in Figure 4.7. It would be an unusual circumstance to underestimate the cost by this amount, but it is possible.

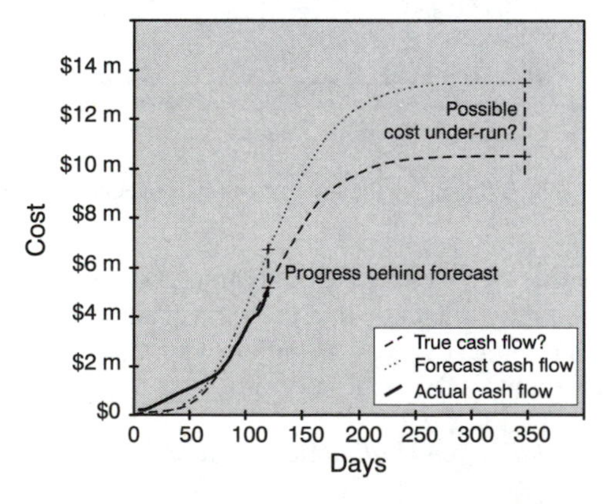

Figure 4.7 Correcting by changing the forecast cost (adapted from Kenley, 2001)

This assessment is more likely when monitoring the contractor's outward cash flow than the inward cash flow from the client, which is generally contractually constrained.

5. Project cost estimate was inflated

The fifth alternative, that the cost estimate was an error and that the project will necessarily have a cost under-run, is the same as illustrated in Figure 4.7.

6. A combination of these factors

The above interpretations seem obvious. What is less obvious, however, is that these same factors can be in operation simultaneously. For example, at a given stage on a project, the project could be over budget but ahead of schedule—thus giving the appearance of being under control. It is therefore very important to properly investigate the project before taking action.

Progress ahead of schedule

The same logic applies for the reverse situation, where the project is ahead of the schedule, although this is less likely to cause concern on most construction projects. The project might be finishing early (Figure 4.8) or it might be suffering a cost blow-out (Figure 4.9).

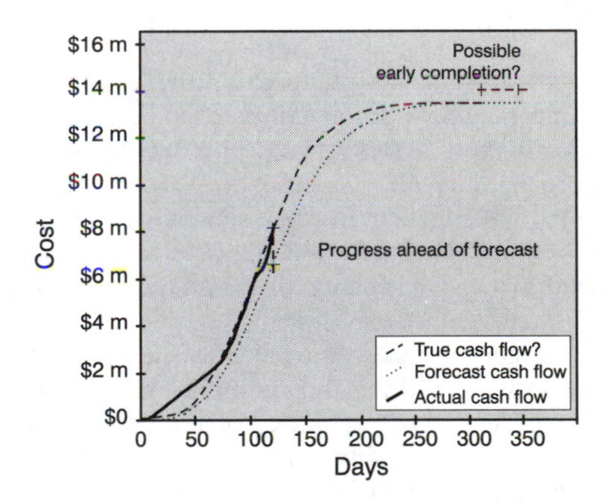

Figure 4.8 The reverse example: project appears ahead of schedule (adapted from Kenley, 2001)

The explanation that the project might finish early is generally the more attractive one and most site managers would tend toward drawing this conclusion.

The alternative explanation, that cost will blow-out, is of much greater concern. The natural tendency to identify with an early project may obfuscate the more serious problem that project cost will run beyond the budget. The interpretation would depend on the type of contract and whether it is a cost or value curve. Such a situation on a cost–plus contract should cause great concern for a client.

Once again, a combination of causes may be at work. Further investigation is required to resolve the cause of the situation.

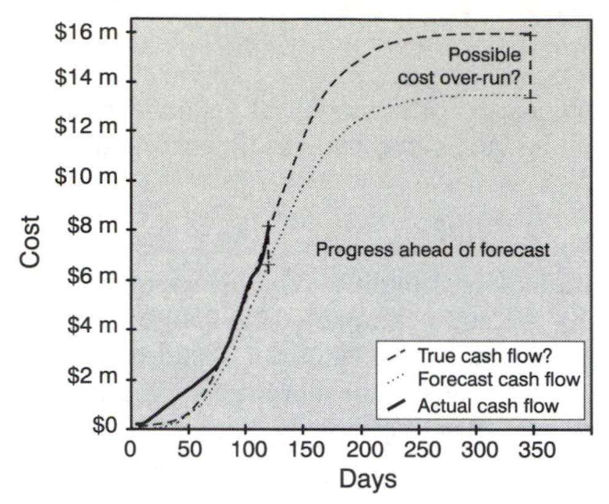

Figure 4.9 The reverse example: project appears cost will over-run (adapted from kenley, 2001)

REMOTE MONITORING OR AUTOMATED SYSTEMS

Forecast and actual cash flow

Integrated and remote management systems can provide an early warning system to identify problems which may not be apparent by examining either a time schedule or cost budget alone. While these systems should be directly applied within the daily management of the project, it is fair to say that their acceptance in this market is limited. Those with a direct involvement in work schedules do not want the extra work involved. Nor do they readily see that they learn anything they didn't already know. In fact it is often difficult to achieve separate time and cost control cycles, without the added complexity of integrated systems.

This situation changes for those removed from the daily management. They are more able to take an overview position and an interest in the project as a whole. 'Top management, as part of their control duties, often rely on the progress reports produced by site managers on the individual projects' (Kaka, 1999). Further afield, the client or client's project manager is not able to participate in the daily management of the site, nor can they always obtain the required information. Very often they must rely on less direct methods. Integrated systems provide valuable information and, more importantly, highlight problems which may not be apparent on the site.

For those operating at arms length from the site, early warning systems become important. It is often too late to act once evidence from time schedule systems becomes clear. The contractor's desire to report that progress is on schedule can delay the identification of problems.

Contractors will generally be keen to advise a client that their project is on schedule during construction. Without access to detailed project information, it is difficult for a client to develop a true picture of project circumstances.

It is worth considering the reality of the in-progress monitoring of the cash flow between a client and a contractor. In the instance of a lump-sum contract, the expected final cost is generally (subject to adjustment) a known figure. Thus, while it might be increasing or decreasing during the project as variations are requested, such variations can be anticipated in the forecast.

Table 4.1 Sample project progress claim forecast

Date	Claim
10th April 1992	Start
31st May 1992	$180,000
30th June 1992	$339,000
31st July 1992	$470,000
31st August 1992	$500,000
30th September 1992	$390,000
31st October 1992	$225,000
31st November 1992	$110,000
31st December 1992	$28,500

This then allows a remote cash flow monitoring system to treat the final cost as a known figure, leaving variation in the rate of cash flow to impact solely on the final duration of the project. This is best argued by way of an hypothetical project. Table 4.1 shows the forecast cash flow for a sample project, charted in Figure 4.10.

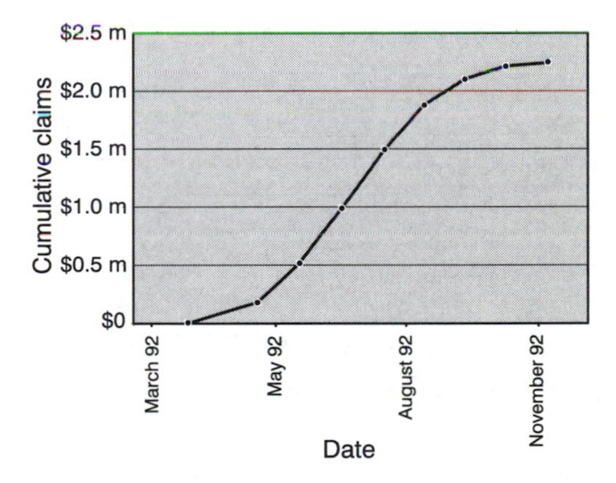

Figure 4.10 Builder's forecast claims for the sample project

If we assume that the project is systematically delayed, then the table of cash flows would be something like Table 4.2, charted in Figure 4.11.

Here it can be seen that the project has blown out in duration, having extended an additional six months from 31st December to 30th June the following year. The data curve is not smooth, but the trend (shape) is similar to the forecast. The distinction between the data and the fitted trend is an important concept.

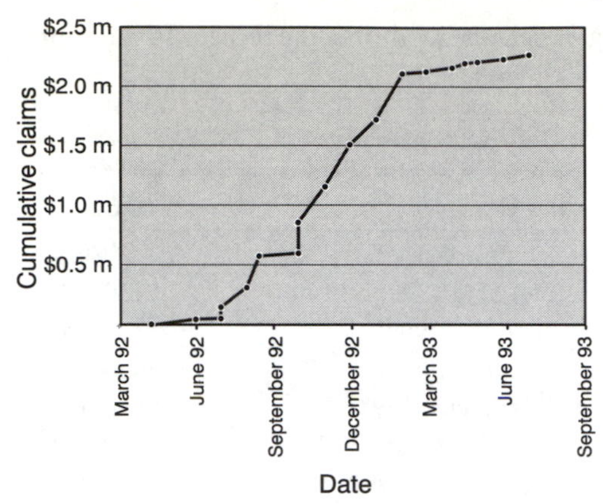

Figure 4.11 Builder's actual claims for the sample project

Table 4.2 Sample project progress claim forecast with systematic delay

Date	Claim
10th April 1992	Start
31st May 1992	\$57,966
30th June 1992	\$97,441
31st July 1992	\$169,568
15th August 1992	\$256,684
30th September 1992	\$278,218
31st October 1992	\$301,436
21st November 1992	\$350,120
31st December 1992	\$209,435
31st January 1993	\$386,515
28th February 1993	\$18,688
31st March 1993	\$30,811
15th April 1993	\$37,705
30th April 1993	\$10,310
31st May 1993	\$24,211
30th June 1993	\$35,653

Using the Logit cash flow model (Kenley and Wilson, 1986), it is possible to fit a trend line to both the actuals and the estimates. For this sample project, the trend line for the actuals makes a very good fit of the forecast data when 'stretched'. Stretching means that a curve of the original profile can be expanded or shrunk proportionally to fit the new situation, with the α and β values remaining the same. The trend is shown in Figure 4.12, with the illusion of stretching highlighted and revealed by the very similar α and β values tabled in Table 4.3.

Table 4.3 Sample project

Data set	α	β	SDY
Estimate	0.3823	1.9591	0.30%
Actuals	0.3201	1.9065	4.40%

Forecasting end dates

The ability to stretch the forecast profile to fit the rate of actual progress may be used to develop a method for forecasting the end date of a project. Figure 4.12 shows the underlying principle that the profile was forecast correctly but the project duration was incorrectly estimated. This is difficult to assess during a project when the actual end date remains unknown, but is an important estimate of project performance. A tool which could assist in forecasting the end date of the project using remote data would be a valuable tool.

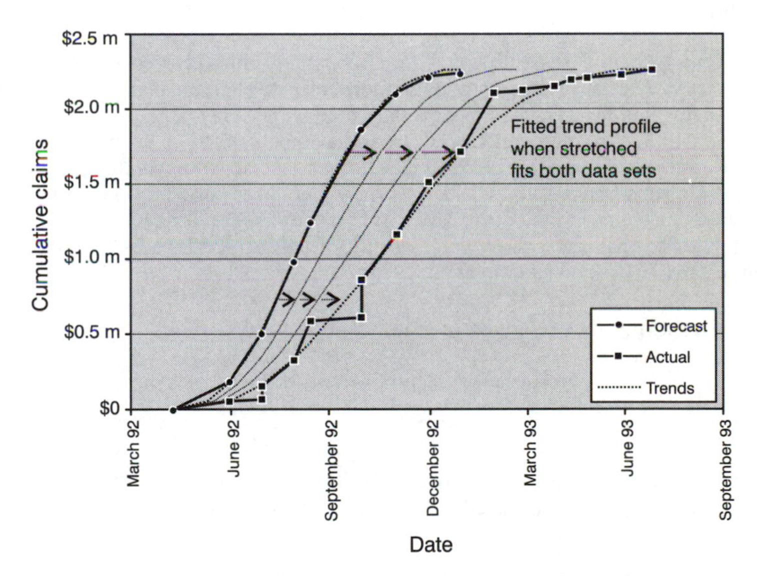

Figure 4.12 Forecast and actual data for the hypothetical project with fitted trend curves. Illustrates the way the forecast trend line fits the actuals when stretched laterally (adapted from Kenley, 2001)

The above result highlights the potential to forecast delays, assuming two circumstances hold true: that the original profile was the correct *shape* for the project, and that the final cost has either not changed or is known. Under these assumptions, it is a simple matter to use the latest progress claim amount and its date to forecast the final amount and date.

There have been relatively few models developed for forecasting the end-date of a project during its progress. However, the DHSS cash flow models included this functionality as a property of their model (Hudson and Maunick, 1974). This was subsequently adopted by those who followed their work (Berny and Howes, 1982), sometimes almost as a passing comment on the nature of the cash flow model but more recently as a specific effort to generate models for forecasting the end date (Kaka, 1999). This issue is discussed further in Chapter 5.

The DHSS standard curve approach

The relationship between the forecast and the actual data as illustrated in Figure 4.12 was always a component of the DHSS model, as first published by Hudson and Maunick (1974). This component of their work was essentially idiographic in nature, but has since been largely ignored by the chief proponents of an idiographic approach Kaka (1999) and Kenley (2001). This neglect is unfortunate, as true credit for developing the method has not been properly awarded. The DHSS model was relatively straightforward, but relied on the use of a standard profile. It is likely that derivation of the nomothetic standard curve which was the aim of their work led to the method being overlooked by many.

> The contract period originally stated for a scheme is seldom correct to the month. Very few schemes reach completion sooner than expected, the majority anything from a few months to several months later than expected.
>
> As the scheme gets underway, the actual expenditure can be plotted on the same graph as the standard curve, and as long as the actual expenditure follows the line of this curve we may take the original contract period as being the best estimate of actual duration available. When, however, the actual expenditure begins to be consistently above or below the normal curve, it is necessary to revise this estimate (Hudson and Maunick, 1974).

The DHSS model was given in Chapter 3 as follows:

$$y=\frac{v}{S}=x+cx^2-cx-\frac{(6x^3-9x^2+3x)}{k} \qquad \text{(3.28 repeated)}$$

where x is the ratio of time given by month m divided by total contract period P and v is the actual expenditure (pounds) and the ratio of expenditure is given by y.

$$x=\frac{m}{P} \qquad (4.1)$$

Substituting this into the equation gives

$$v=S\left[\frac{m}{P}+c\left(\frac{m}{P}\right)^2-c\frac{m}{P}-\frac{\left(6\left(\frac{m}{P}\right)^3-9\left(\frac{m}{P}\right)^2+3\frac{m}{P}\right)}{k}\right] \qquad (4.2)$$

which may be transformed to solve for P (Hudson, 1978):

$$\frac{kvP^3}{S}+(ck-k+3)mP^2-(ck+9)m^2P+6m^3=0 \tag{4.3}$$

For example, assuming that the project forecast was based on the DHSS profile for a £6 million project giving $C = 0.192$ and $k = 3.458$, then a project of value £6.1 million and expected to take 48 months could be expected to have reached after 28 months (58.33%) a value of 63.95% or £3,839,756. If, however, expenditure has only reached £3.5 million, then solving the above equation yields $P = 20, 48$ (total duration will be 48 months) or at £3 million then solving the above equation yields $P = 20, 53$ (total duration will be 53 months). There are usually two solutions to the cubic Equation 4.3 within the subject range. The solution curve is shown in Figure 4.13.

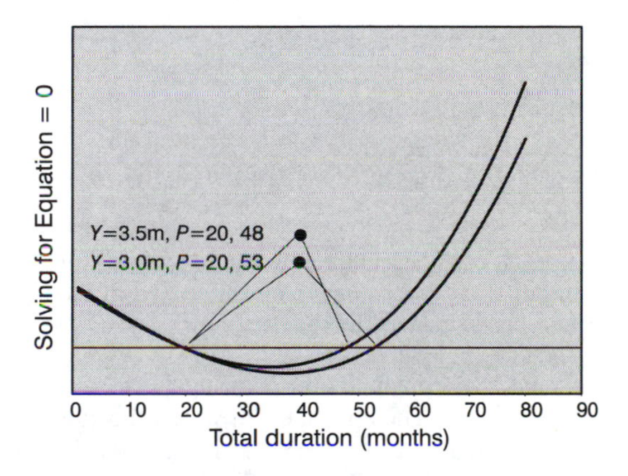

Figure 4.13 Solving for Equation 4.3 = 0

Berny and Howes

The Berny and Howes (1982) formula was based on the DHSS model. They did not publish an equivalent equation to solve for P and seem rather to have relied on graphical means. However, an equivalent calculation can be made for their two-parameter formula.

The Berny and Howes cash flow formula was:

$$y=\frac{v}{S}=x\big(1+a(1-x)(x-b)\big) \tag{3.29 repeated}$$

$$y=-ax^3+(a+ab)x^2+(1-ab)x \tag{3.30 repeated}$$

which may be transformed to solve for P:

$$a-(a+ab)\frac{P}{m}-(1-ab)\left(\frac{P}{m}\right)^2+\frac{v}{S}\left(\frac{P}{m}\right)^3=0 \quad (4.4)$$

This achieves identical results to the DHSS Equation 4.3, with curves matching Figure 4.13 although on a different scale. Thus using the Berny and Howes model with $a = 1.7351$ and $b = 0.6107$, $P = 20, 48$ for $v = £3.5$ m and $P = 20, 53$ for $v= £3$ m. Thus either model can be used to calculate the likely end date when using the DHSS standard profiles.

Berny and Howes considered that five months were required before an estimate of final end-date could be made.

The DHSS and Berny and Howes models are difficult to use and, as they are cubic, they require either user intervention and iteration, or solving by simultaneous equations. This is neither user friendly nor readily incorporated into an automated remote system.

Kaka's stochastic benchmark model

Kaka's review of the cash flow literature confirmed, in his view, that the idiographic approach is valid, that:

> Construction projects are unique and future attempts to standardise the cost/value flow are likely to fail. Therefore, a new concept or type of model is needed to replace these deterministic models (Kaka, 1999).

Kaka adopted a stochastic approach to monitoring performance on construction projects using cash flow curves. The approach involved the stochastic generation of an envelope of cost flow curves based on the historic data. He drew upon the monthly cost commitment curves of 118 projects, collected from four contractors in an innovative approach designed to address the problem of the failure of standard models to achieve a good forecast for construction project cash flow. He divided projects into categories by duration (up to seven months, seven to 12 months and longer than 12 months) then calculated α and β for each project, using the Kenley and Wilson (1986) model.

$$v=\frac{100e^{\alpha}\left(\frac{t}{100-t}\right)^{\beta}}{1+e^{\alpha}\left(\frac{t}{100-t}\right)^{\beta}} \quad \text{(3.5 repeated)}$$

$$\text{or } v=\frac{100F}{1+F} \text{ where } F=e^{\alpha}\left(\frac{t}{100-t}\right)^{\beta} \quad \text{(3.6 repeated)}$$

The α and β values for each project allowed the calculation of values at time t=30%, 50%, 70%. From this he derived averages at those points with the calculated standard deviations. Kaka's control system:

> relies on the calculation of the expected duration which is essentially an update of the duration forecast based on current performance. In the proposed stochastic model a range of possible durations would be forecast and the project manager uses this to assess the probability of meeting the contracted duration (kaka, 1999).

His model did not use the averages directly, however, as Kaka wished to model the individual nature of projects. He therefore derived the values for t=30% and t=70% from t=50% using the relationship between these values, which he identified to be strongly positively linear.

Table 4.4 Kaka's (1999) regression of cost between t = 50% and t = 30%, 70%

		Contract's duration falling within		
		0–6 months	7–12 months	13–36 months
t = 50% Mean		74.150	55.797	45.517
t = 50% Standard deviation		14.961	8.953	8.033
from t = 30% to t = 50%	Regression	$y = 1.012x - 28.44$	$y = 0.63x - 6.06$	$y = 0.562x - 5.03$
	Coefficient of determination (r^2)	0.80	0.58	0.73
	Error of fit (Standard deviation)	7.623%	4.803%	2.78%
from t = 50% *to* t = 70%	Regression	$y = 0.53x + 51.30$	$y = 0.723x + 38.9$	$y = 0.87x + 33.07$
	Coefficient of determination (r^2)	0.83	0.78	0.83
	Error of fit (Standard deviation)	3.572%	3.416%	3.172%

The relationship between cost at t=50% and t=30%,70% for each of the project groupings is given in Table 4.4. Kaka provided r^2, described as the *goodness of fit*. As the results were charted by a spreadsheet program it is reasonable to assume that these were calculated in Microsoft Excel. Microsoft provide the following equation for the Coefficient of Determination r^2:

$$r^2 = 1 - \frac{SSE}{SST} \tag{4.5}$$

where $SSE = \sum(Y_j - \hat{Y}_j)^2$ and $SST = (\sum Y_j^{\ 2}) - \frac{(\sum Y_j^{\ 2})}{n}$

and $\hat{Y}$ is the fitted estimate of Y.

To derive the envelope of profiles, Kaka started with values at 50% duration using the mean and standard deviation. From this the values at t =30%, 70% were derived using the linear relationship and the error of fit (Table 4.4). This yielded three points in each iteration, enabling the calculation of α and β for each simulated project. The resultant α and β values were not published.

From a given set of α and β values, Kaka was able to calculate a range of expected completion times D for a given actual elapsed time t and cost c:

$$D=\frac{tc}{t}=\frac{e^x}{1+e^x} \text{ where } x=\left(Log_e\left(\frac{c}{1-c}-\alpha\right)\right)\frac{1}{\beta} \quad (4.6)$$

where c is % current cost to total cost, and α and β are the envelope curve parameters.

Kaka's model provides a range of reasonable outcomes, based on past analysis, within which a project can be expected to finish. It yields the probability of project duration running over the targeted end date. However, this is based an envelope of possible cash flow profiles and a single data point. The problem with this approach is that there is no way of knowing, from the single data point, where that point sits with regard to the general progress of the project.

The model provides a powerful mechanism for providing early warning systems for risks of time over-run. Should the project data points move into the area where there becomes a strong probability of time over-run, then management could take remedial action.

The Logit in-project end-date forecast model

The underlying principle of the Kaka model may be extended. Using the logic followed by Hudson and Maunick (1974), Berny and Howes (1982) and graphically represented in Figure 4.12, it is possible to use the Kenley and Wilson (1986) Logit model to forecast the likely end date for a project (Kenley, 2001).

The Hudson and Maunick/Berny and Howes approach was to use standard curves. This is one way to provide a forecast cash flow. The Kaka model rejected this approach because of the inability to derive an accurate forecast and instead used a stochastic approach. Each of these methods requires the forecast of a cash flow for a project. The following method accepts that there are limitations in project forecasts but nevertheless recognises their role in a contract. The assumption is that if a contractor has provided a value curve (inherent in a claims forecast) then they expect it to match the real claims profile for the project. In the event that the actual claims profile varies from the forecast, then there are reasonable grounds for concern. The same could apply for site management, where the cost profile was being monitored.

The method required the development of an extension to the Logit cash flow model to allow the derivation of time given cost. The model, outlined in Equations 3.1 to 3.9, is not suitable for deriving an estimated time at which a certain value would be reached. The following illustrates an alternative model whereby time is expressed as a function of cost using the Logit model.

The logistic equation for cash flows is normally expressed using value v as the dependent variable and time t as the independent variable (Equation 3.2). This may however be re-expressed using time t as the dependent variable and value v as the independent variable (Equation 4.7)

$$\ln\left(\frac{v}{1-v}\right)=\alpha+\beta\left(\frac{t}{1-t}\right) \qquad \text{(3.2 repeated)}$$

$$\beta\log_e\left(\frac{t}{1-t}\right)=\left(\frac{v}{1-v}\right)-\alpha \qquad (4.7)$$

This then forms the equation of the sigmoid curve which describes the flow of cash on a specific building project. It may also be expressed in terms of v as follows:

$$t=\frac{\left(e^{-\alpha}\left(\frac{v}{1-v}\right)\right)^{\frac{1}{\beta}}}{1+\left(e^{-\alpha}\left(\frac{v}{1-v}\right)\right)^{\frac{1}{\beta}}} \qquad (4.8)$$

$$\text{or } t=\frac{G}{1+G} \text{ where } G=\left(e^{-\alpha}\left(\frac{v}{1-v}\right)\right)^{\frac{1}{\beta}} \qquad (4.9)$$

The Logit cash flow model given above uses scales from 0.0 to 1.0, where the ratio (on the abscissa or ordinate) 1.0 is equivalent to 100%. Where percentage scales are to be used in accordance with convention, the equations should be expressed as follows

$$\text{If } \beta\ln\left(\frac{t}{100-t}\right)=\left(\frac{v}{100-v}\right)-\alpha$$

$$\text{then } t=\frac{100\left(e^{-\alpha}\left(\frac{v}{100-v}\right)\right)^{\frac{1}{\beta}}}{1+\left(e^{-\alpha}\left(\frac{v}{100-v}\right)\right)^{\frac{1}{\beta}}} \qquad (4.10)$$

$$\text{or } t=\frac{100G}{1+G} \text{ where } G=\left(e^{-\alpha}\left(\frac{v}{100-v}\right)\right)^{\frac{1}{\beta}} \qquad (4.11)$$

Equation 4.11 looks similar to the Kaka equation given in 4.5, but is consistent with the cash flow equations presented elsewhere and reduces the equation to simpler and less confusing terms. Equation 4.11 is the inverse function of Equation 3.6.

Equation 4.11 can be used to solve for time given a value. If the α and β values are known for the project forecast (they may be calculated for a table of forecast progress claims) then the planned time for reaching that value may be calculated. It was demonstrated in Figure 4.12 that the *stretching* effect holds (in principle) through the life of the project. In order to forecast part way through the project, it is only necessary to take the latest cumulative amount.

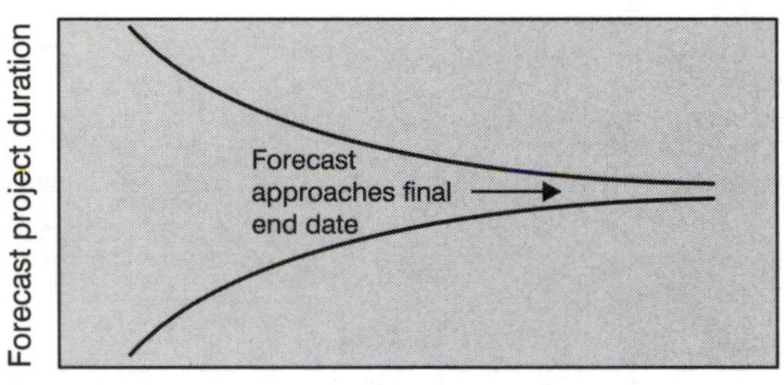

Figure 4.14 Diagrammatic representation of variability in the forecast, of the end date, as the project progresses (adapted from Kenley, 2001)

The calculation of the revised end date uses the original forecast cash flow for the project as the seed (Kaka's stochastic method would be an alternative approach, or a standard curve profile), and to tabulate the results. This produces a fluctuating forecast which can be expected to approach the actual final end-date as the project proceeds. Diagrammatically, the forecast of the final end date made during the life of the project is shown in Figure 4.14.

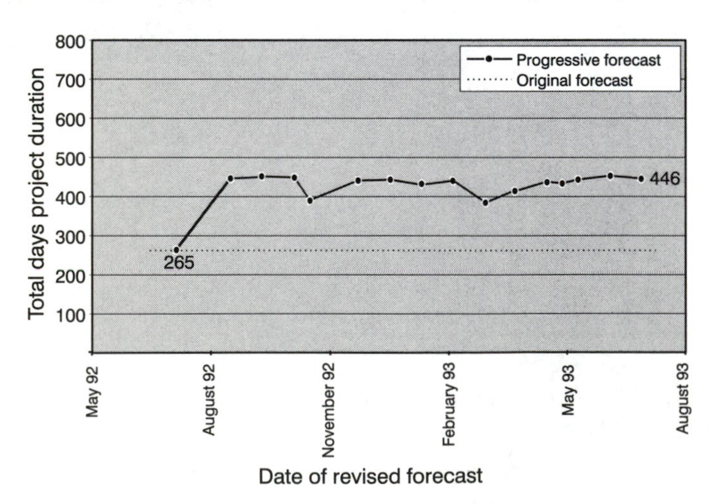

Figure 4.15 Actual forecast end-date during the progress of the project (Kenley, 2001)

This model may be demonstrated with the sample project used in Tables 4.1 and 4.2 and Figures 4.10 and 4.11, where the progressive forecast end date for the project is tabulated in Table 4.5 and illustrated in Figure 4.15.

In Figure 4.15 the forecast profile (shape) is accurate but the end date originally forecast was not. This became apparent very early on in the project.

Where the project costs and schedule closely follow the profile of the forecast, then there can be a high expectation of accuracy for the progressive forecast end date. Where the forecast is not accurate (as suggested by Kaka) then the project will approach the final end-date as the project progresses.

Table 4.5 Progressive forecast of final duration upon receipt of claim

Date	Claim	Forecast duration
10th April 1992	Start	265
31st May 1992	$57,966	446
30th June 1992	$97,441	452
31st July 1992	$169,568	449
15th August 1992	$256,684	391
30th September 1992	$278,218	441
31st October 1992	$301,436	443
21st November 1992	$350,120	430
31st December 1992	$209,435	440
31st January 1993	$386,515	385
28th February 1993	$18,688	413
31st March 1993	$30,811	438
15th April 1993	$37,705	433
30 April 1993	$10,310	443
31st May 1993	$24,211	453
30th June 1993	$35,653	446

This model is only limited in its accuracy by the relevance of the forecast to the final project. It is assumed that the final contract sum is known during the project. Clearly this is subject to adjustment due to variations during the project, however it is assumed that the latest adjusted contract sum would be used for the analysis.

In summary, the steps in the model to progressively forecast the end date for a project are:

- The Logit α and β values are calculated for the forecast cash flow profile.
- The forecast final completion date is progressively adjusted for changes in the project 'lump-sum' through variations. For the post hoc analysis of projects, the difference between the original and the final contract sum may be applied progressively.
- The forecast time for that value is calculated, using the revised project duration and lump-sum.

- The ratio of forecast to actual is calculated.
- This ratio is then applied to the duration to establish a revised 'stretched' duration, from which an end date may be calculated.

Forecasting real projects

Real projects can be very different. First, as Kaka stated, it is hard to predict a project cash flow. Secondly, estimators are usually under extreme pressure at tender time, and forecasting cash flow must seem irrelevant to the task at hand. This understandable lack of priority reduces the confidence in the forecast.

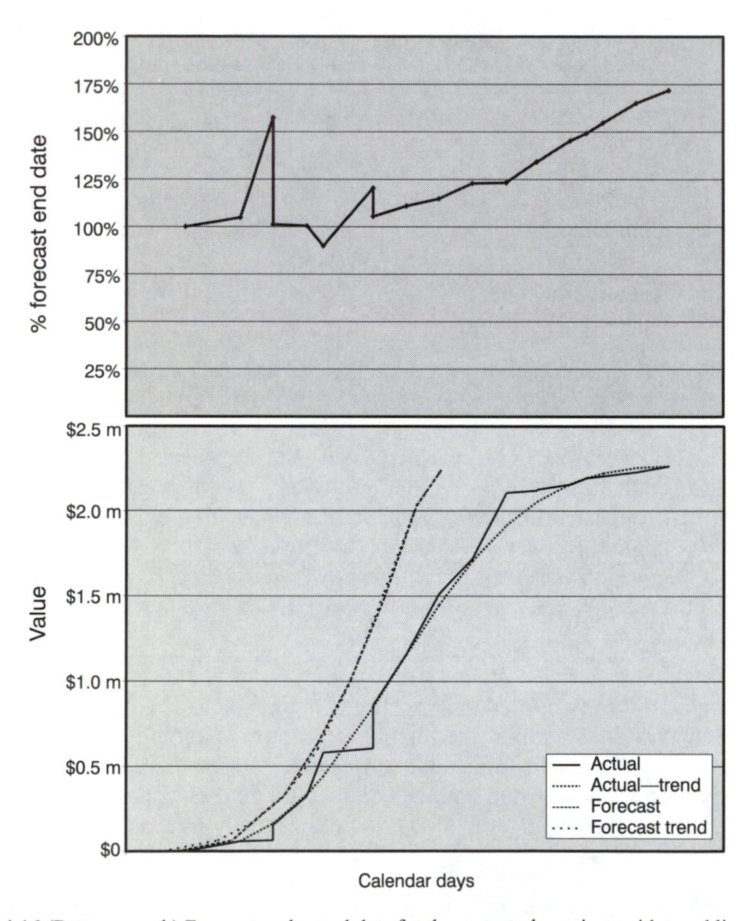

Figure 4.16 (Bottom graph) Forecast and actual data for the case study project, with trend lines fitted. (Top graph) Corresponding progressive forecast end date (Kenley, 2001)

To test the model under real conditions, twenty-two projects were tested from a range of contexts, including 21 Australian public sector projects and one Malaysian private sector project. There were two subsets: a set of similar projects (new fire stations on green-field sites); and a further set of dissimilar projects from very different contexts. There were eight new fire stations, six police stations, five educational facilities, a major commercial refurbishment, an industrial building and the Malaysian project was a major housing development (conglomerate structures, not individual dwellings) in Kuala Lumpur.

One of the projects from the second data set is selected as a case study to illustrate the results. This case study is the project on which the sample project from the previous section was based. The project is used to illustrate the model because it is a good example of a project consistently behind schedule. The demonstration project was commenced in winter, in Melbourne, Australia. It was an industrial building built on highly-reactive swamp land. The year of its construction was one of the wettest on record in Melbourne, raining almost every day for months on end, and this severely delayed the project. The builder was allowed extensions of time, but no associated costs. As a result, the builder wanted to believe, and continued to report, that they would complete the project on time, despite the delays. Monitoring the schedule indicated this was unlikely, but the schedule was adjusted by the contractor to indicate maintenance of the original completion date. Certainly the builder tried very hard in very difficult circumstances, however, it was only by use of the above model that the true position was indicated to the client. The cash flow profile told a different story from that being disclosed at site meetings. This project was an example of the use of this technique as a remote management tool by the client's project manager, and demonstrates that financial data alone can highlight problems with progress.

Table 4.1 provided the original forecast of progress claims provided by the builder. The contrasting actual figures have already been provided in Table 4.2. Figure 4.16 shows (a) the component cash flow curves and data, and (b) the forecast end date.

Results

A sample of project cash flow charts and their resultant end forecast are shown in Figures 4.17 (1) to (3). These illustrate typical results from the data set. Here Figure 4.17 (1) is an educational facility in a regional Victorian town, Figure 4.17 (2) is the refurbishment of a significant public building in Melbourne and Figure 4.17 (3) is a large-scale housing project in Kuala Lumpur, Malaysia. The following may be noted for the three projects in Figure 4.17:

- Figure 4.17 (1) shows a systematic delay in the works which is predicted consistently through the project.
- Figure 4.17 (2) shows an increase in project value, corresponding duration, but also consistent prediction of a time blow-out.
- Figure 4.17 (3) shows the close matching of the estimate with actuals, with a corresponding consistent prediction of attaining the desired end date.

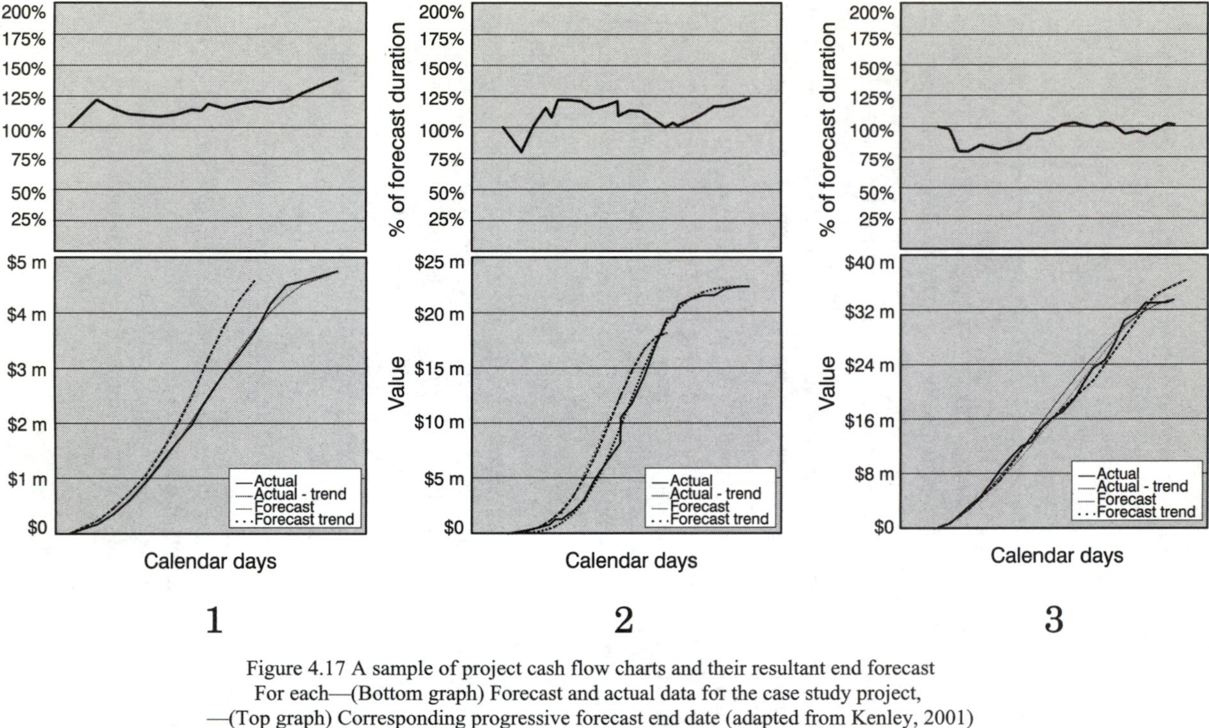

Figure 4.17 A sample of project cash flow charts and their resultant end forecast
For each—(Bottom graph) Forecast and actual data for the case study project,
—(Top graph) Corresponding progressive forecast end date (adapted from Kenley, 2001)

The collected progressive forecasts for all projects within the two samples are illustrated in Figures 4.18 and 4.19. The following were observed in the projects analysed:

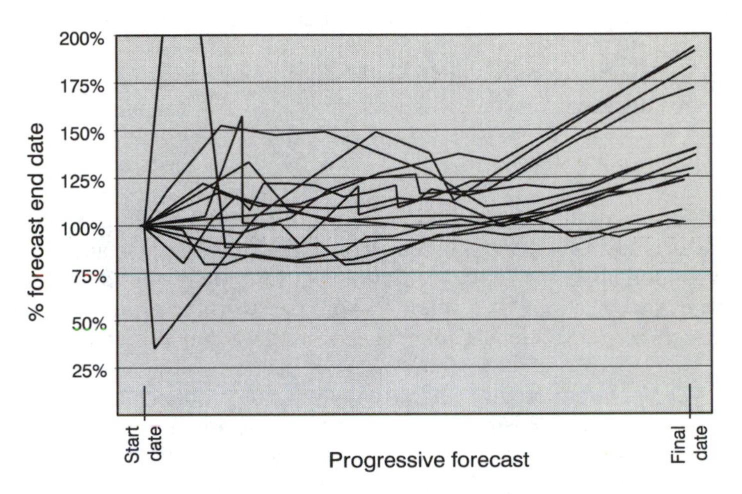

Figure 4.18 Progressive forecasting of end date, for various major projects (Kenley, 2001)

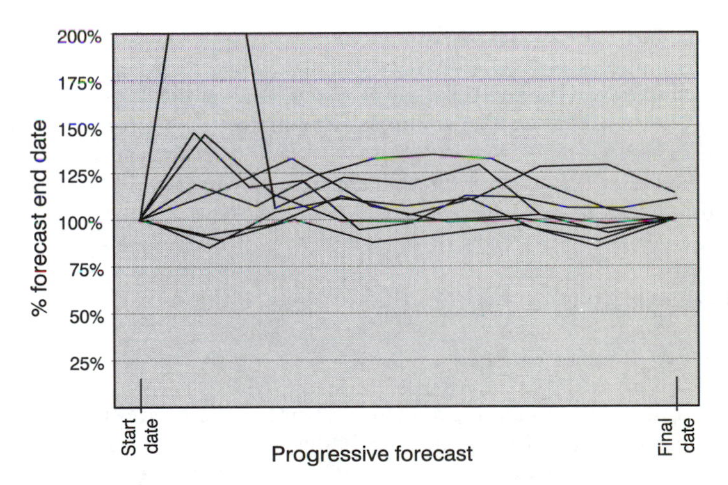

Figure 4.19 Progressive forecasting of end date, for a set of small institutional buildings (Kenley, 2001)

- There was a tendency to use a cash flow forecast which did not include a slow-down period. This resulted in a frequent duration blow-out in the latter stages of the projects, therefore not predicted by the model. This demonstrated a systematic error on the part of estimators.
- Projects systematically failed to achieve the forecast end date. Only those projects with an early completion (<100%) forecast early and progressively tended to approach the expected end-date (100%) at the end of the project.

- There was a frequent blowout in the very last payments, due to problems finalising the project. These were generally for small sums, usually not related to practical completion, and thus not important for the predictive model. However, clients might well recognise the likelihood of problems closing out projects which these results reflect.
- Smaller projects were more likely to finish close to the adjusted final time and were more consistent.

Conclusions

The aim in designing this method was to achieve a model for in-progress end-date forecasting as a means of providing remote monitoring of performance on projects.

The method appears to have great potential for monitoring the progress of the works during construction. It does not rely on access to the project schedule or its updates and therefore is suitable for inclusion in an automated financial reporting system.

The system would provide an early warning of problems and would thus be suitable as a control mechanism. However, it must be recognised that any control system necessarily effects the project data. Early advice of problems does not of course dictate continuance of those problems, due to feedback mechanisms.

INTEGRATIVE TECHNIQUES

The problem of resolving the cause of problems on a project, or to develop early warning of problems, has another solution. Through the use of project schedules, it is possible to interrogate the progress of the work and the value of the work done, to develop a picture of the progress through the 'earned value' which can be compared against project targets.

Earned value analysis

Introduction to earned value analysis

Earned value is an integrated method for measuring the performance of a project using both the schedule of work, the cost budget and the progress of work against these two plans. This method is a powerful aid in solving the questions discussed previously, about deviation of the cash flow from the planned.

Earned value analysis has risen from systems developed for the administration of progress payments for major capital works purchases. As discussed in Chapter 3, progress claims for this type of capital works are in effect a loan to the contractor to assist them to afford to complete the works. Under this situation it is very important to ensure that payments do not get ahead of the progress of the works (payment in advance) and it is equally important to monitor the progress of the works completed against the budget. These systems have previously been known as Cost/Schedule Control Systems Criteria (C/SCSC), as developed by the USA Department of Defence and in use since 1967 but renamed Earned Value Analysis (EVA) in 1996.

The tool was designed by managers who recognized that increasing program complexity and proliferating management systems demanded a reasonable degree of standardisation. Thus it is both a performance management system and a method of standardisation of the administration process.

There is considerable richness to the earned value method, but it is only the performance measurement functionality which is of concern here. For performance management, the 'earned value' method is also known as 'achieved value', 'accomplished value', 'physical quantity measurement' and 'Earned Value Performance Management' (EVPM).

Fully implemented, earned value requires segmenting a project into controllable parts using a 'work breakdown structure' (WBS) which is related to the 'organisational breakdown structure' (OBS). The WBS includes all work tasks for the project. Costs are similarly broken down into a 'cost breakdown structure' (CBS). Bent and Humphreys (1996) note that the WBS is in fact a subset of the CBS—which extends to include those cost items which do not involve work.

At the heart of the breakdown, there are cost accounts consisting of either labour or cost, and a direct relationship is established between the percentage of work done and the budget for that account.

Earned value performance management utilises both mathematical and graphical relationships.

Mathematics of earned value analysis

The performance of the project is measured through a series of indicators, with the earned value (EV) being for each cost account:

$$EV = (Percentage\ complete) \times (Budget\ for\ the\ account) \quad (4.12)$$

The summing of these provides the EV for the project—which is equivalent to the work in progress curve. Performance measurement, however, requires this integration of time reporting and cash expended implicit in Equation 4.12. Furthermore, performance measurement requires comparison of the rate of progress with the originally planned rate of progress, or base-line scheduled progress.

Progress variance has been established to be either a variance in the time performance—'schedule variance' (*SV*) or a variance in the cost performance—'cost variance' (*CV*), or both. These indicators are:

$$SV = BCWP - BCWS \quad (4.13)$$

$$CV = BCWP - ACWP \quad (4.14)$$

where

$$BCWP = Actual\ cost\ of\ work\ performed = EV \quad (4.15)$$

$$BCWS = Budgeted\ cost\ of\ work\ scheduled \quad (4.16)$$

$$ACWP = Actual\ cost\ of\ work\ performed \quad (4.17)$$

These measures provide a rapid feedback mechanism.

If *SV* is positive, then the project has completed more work than scheduled by cost (a different answer may be possible were performance by quantity to be measured). If *SV* is negative, then the project is behind schedule.

If *CV* is positive, then the project is costing less than budgeted. If *CV* is negative then the actual cost has exceeded the budgeted cost for the work performed.

Together these two indicators inform about both schedule and cost performance. It is also possible to develop integrated measures of efficiency. These are 'cost performance indicator' (*CPI*) and 'schedule performance indicator' (*SPI*):

$$CPI = BCWP/ACWP \tag{4.18}$$

$$SPI = BCWP/BCWS \tag{4.19}$$

An index value of 1.0 or greater indicates better than planned performance for either indicator. A value less than 1.0 indicates poor performance relative to the plan.

Barr (cited in Meredith and Mantel 2001) provides a further indicator, 'Cost–Schedule index' (*CSI*) which combines *CPI* and *SPI*:

$$CSI = CPI \,.\, SPI \tag{4.20}$$

CSI and resolves the situation where one of the ratio indicators is less than 1.0 and the other is greater than 1.0. A problem is indicated where $CSI < 1.0$.

Applying earned value analysis

Earned value analysis requires sophisticated software to be able to track the costs and progress for each individual cost centre.

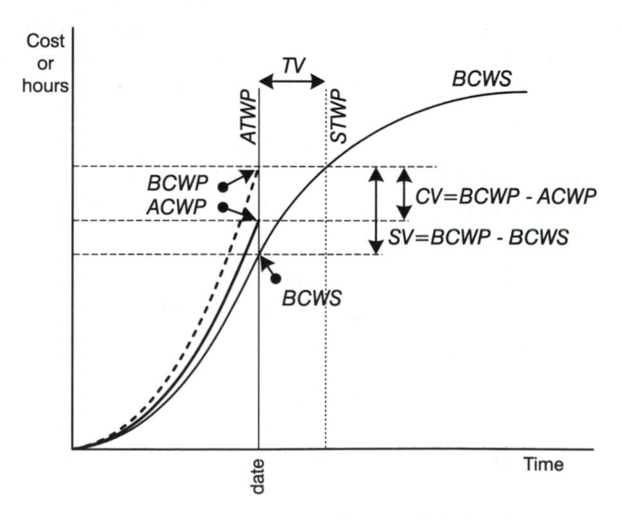

Figure 4.20 Earned value analysis—case 1

There are now many software packages which will undertake this task. However, there may be a difficulty obtaining cost data in a timely manner as accounts departments usually operate on different systems and timelines, resulting in difficulty integrating this data. Therefore, many analysts use labour and man-hours to track earned value. This is a relevant approach where the bulk of the cost and the variability in cost is directly proportional to the hours expended. This may not be suitable where material consumption or waste is a significant and variable factor.

Figure 4.20 illustrates a case where the project appears ahead of schedule and is both saving money and ahead of schedule. Time variance (*TV*) is read graphically as the time difference between *BCWS* and *BCWP*, which is the difference between scheduled time of work performed (*STWP*) and the actual time of work performed (*ATWP*).

In contrast, Figure 4.21 illustrates a case where the project appears ahead of schedule but is both losing money and behind schedule. This clearly demonstrates the power of EVPM.

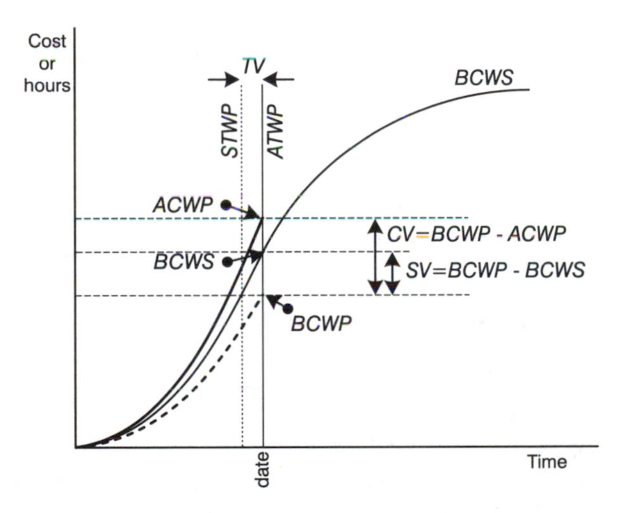

Figure 4.21 Earned value analysis—case 2

The stochastic approach to EV

Barraza *et al.* (2000) constructed a stochastic model integrating the time schedule and the budget. This was a graphical mechanism for monitoring progress, closely related to the earned value method upon which it was based. This is an extension of the model discussed by Bent and Humphreys (1996: 314), but inverted to express time as a function of cost.

These models should assume a sigmoid relationship exists between the stochastic distribution of time with value and chart an expanding 'tube' of possible s-curves. When displaying the relationship most authors display a linear or even convex relationship which rapidly moves to full variability early in the project. If the probability distribution is to hold, then the curve would be a cumulative distribution of the normal distribution as shown in Figure 4.22. This results in a 'tube' of potential cash flow which is quite considerably more constrained, moving slowly to full variability later in the project, and contrasts with that indicated by Barraza, *et al. (2000)*.This suggests that very small fluctuations early in the project may indicate larger fluctuations toward the end.

The method is similar to Kaka's (1999) benchmark model in its use of an envelope of acceptable or reasonable outcomes allowing rapid recognition of problem projects, but is based on the schedule rather than on standard profiles.

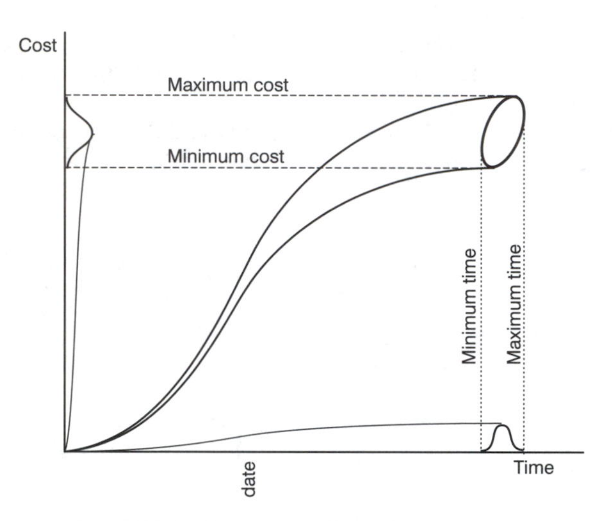

Figure 4.22 Stochastic range of earned value outcomes

CONCLUSION

Performance measurement is critical to the early identification of problems on a project. The suite of time-performance techniques which have been discussed in this chapter allow various ways to monitor progress. These techniques vary from remote and independent monitoring of gross cash flow profiles, to the data-intensive integrated cost-schedule based methods of earned value analysis.

While such methods have been available for a long time, acceptance has been slow in mainstream construction, most likely due to the cost and intensiveness of integrated systems. The much cheaper and more easily automated remote monitoring mechanisms described here, may offer an alternative where full integration is not possible or desired. Ultimately, the development of stochastic approaches from an integrated basis or from a standard curve basis, results in the same use of an envelope of reasonable outcomes.

Further research into the comparison of integrated and remote techniques would be a valuable aid to assist organisations to choose their best monitoring methods.

REFERENCES

Barraza, G. A., Back, W. E. and Mata, F. (2000). 'Probabilistic monitoring of project performance using SS–Curves'. *Journal of Construction Engineering and Management* **126**(2): 142–148.

Bent, J. A. and Humphreys, K. K. (1996). *Effective Project Management through Applied Cost and Schedule Control.* New York, Marcel Dekker.

Berny, J. and Howes, R. (1982). 'Project management control using real time budgeting and forecasting models'. *Construction Papers* **2**(1): 19–40.

Betts, M. and Gunner, J. (1993). 'Cash flow forecasting'. Chapter 11 in *Financial Management of Construction Projects: Cases and Theory in the Pacific Rim*. Singapore, Longman: 207–227.

Bromilow, F. J. and Henderson, J. A. (1977). *Procedures for Reckoning the Performance of Building Contracts*. Highett, Commonwealth Scientific and Industrial Research Organisation, Division of Building Research.

Horner M. and Duff, R. (2001). *More for Less: A Contractor's Guide to Improving Productivity in Construction.* London, Construction Industry Research and Information Association.

Hudson, K. W. (1978). 'DHSS expenditure forecasting method'. *Chartered Surveyor—Building and Quantity surveying Quarterly* **5**(3): 42–45.

Hudson, K. W. and Maunick, J. (1974). *Capital Expenditure Forecasting on Health Building Schemes, or a Proposed Method of Expenditure Forecast.* London, Research Section, Department of Health and Social Security, Surveying Division.

Kaka, A. P. and Price, A. D. F. (1991). 'Relationship between value and duration of construction projects'. *Construction Management and Economics* **9**(4): 383–400.

Kaka, A. P. (1999). 'The development of a benchmark model that uses historical data for monitoring the progress of current construction projects'. *Engineering, Construction and Architectural Management* **6**(3): 256–266.

Kenley, R. (2001). 'In-project end-date forecasting: an idiographic, deterministic approach, using cash flow modelling'. *Journal of Financial Management of Property and Construction* **6**(3): 209–216.

Kenley, R. and Wilson, O. D. (1986). 'A construction project cash flow model—an idiographic approach'. *Construction Management and Economics* **4**: 213–232.

Meredith, R. R. and Mantel, S. J. (2001). *Project Management: A Managerial Approach*, New York, John Wiley and Sons.

Stacey, R. D. (1996). *Strategic Management of Organisational Dynamics*. London, Pitman Publishing.

Wiener, N. (1996). *Cybernetics: or Control and Communication in the Animal and the Machine*. Boston, The MIT Press.

5

The time–cost relationship

INTRODUCTION

Cash flow modelling is essentially about the relationship between time and cost. While the investigation of project S curves is about the rate of change of that relationship during the life of the project, there is another, equally important, element to the relationship between time and cost; the relationship between final time and final cost.

In order to undertake cash flow modelling or forecasting, it is necessary to have both time and cost forecasts. Forecasting construction cost does not require much discussion as it has been addressed in depth elsewhere. Similarly, forecasting construction duration is a well developed field. However, the relationship between them, where one can be derived from the other, is less well understood. Given that cash flow modelling is usually conducted in the absence of detailed project schedules, or even despite them, there is a need to be able to forecast project duration or to check a forecast duration, rapidly and without resorting to detailed scheduling techniques.

In that context, this chapter will discuss tools available for producing these forecasts from the properties of the relationship. There are two main components to this discussion: the ability to derive a relationship between project cost and time for given samples of projects, and the consequential ability to forecast the end date of a given project (or for categories of projects) given cost.

The time–cost relationship has now been explored through a series of studies, over an extensive period of time. Recently, researchers have turned their attention to seeking long-term trends in the relationship between time and value, to develop models that allow the rapid forecast of project duration using only the budgeted cost and current cost indices.

Furthermore, while the relationship between time and cost is now well understood, it is only recently that attention has turned to exploiting this relationship to explore issues beyond the project, such as issues of long-term trends and movements and international comparisons of industry efficiency or productivity. The discussion here is part of a resurgent interest in time–cost modelling, which may inform such critical issues as improvements in productivity, particularly annual targets.

Governments around the world are seeking to improve the productivity of their construction industries, particularly as they generally play such a large role in the gross domestic product of the nation, yet there is little in the way of tools for measuring that productivity improvement. The relationship between time and cost may provide such a tool.

THE TIME–COST RELATIONSHIP

The time-cost relationship directly correlates project value with project time.

Terminology

Project value

Final project value has two alternatives: the original contract sum and the final contract sum. These are necessarily related, yet there is remarkable variance between them. This variance arises from changes to the contract through variations and time-adjustments. Bromilow (1970) found an exponential relationship between project value and the number and value of variations. It may be seen that this approximates the relationship between time and project value. This may indicate a direct relationship between the value of variations and project time.

Project time

Project value also has two primary values, the original contract duration and the final adjusted duration following extensions of time. Bromilow *et al.* (1988) identified that, on average, projects ran over by 32% for government projects and 22% for private projects, which is greater than the earlier apparent result of 17%. This difference between contract duration and actual duration includes allowable delays for weather and thus is consistent with other duration indicators.
Since the late 1960s, many researchers have explored the apparent relationship between the duration of a project and the building price (frequently referred to as the time–cost relationship). Bromilow, in the late 1960s, explored this relationship and derived an equation which is still in use today. Since that time, Australian and international studies have validated the time–price relationship, have categorised types of projects that may have different performance characteristics and have recalculated or updated the constants. Of particular interest is the parameter which describes the average construction time–performance for a sample of projects, as many authors believe that it indicates the construction time–performance of the industry as a whole at any given time. Research into the time–cost relationship has explored the relationship between the final cost and final time, however different results may be obtained from the relationship between original contract cost and time.

Modelling the duration–size relationship

Single factor indicators

The duration–size relationship is technically the correct form of the relationship. Convention has accepted that size and cost are interchangeable as are duration and time. In a seminal paper in the Building Forum, Bromilow (1969) first expressed time as a function of building value in 1969 . Bromilow referred to '...a simple cost–time relationship' to represent the relationship between the size of a project and the time actually taken to construct. The formula he developed for time as a function of cost was given as:

$$T = KC^{B} \tag{5.1}$$

where T was the actual construction time in working days; C was the cost of building in millions of dollars; K was a constant—characteristic of building time performance in Australia and; B was a constant—indicative of the sensitivity of time performance to cost level. The value of K was given as 350. The value of B was given as 0.30 It was noted that when C equals 1 (AU$1 million), then $T=K$ (350 working days), so K represents the average working time for a $1 million project and is often considered to be an efficiency or productivity measure.

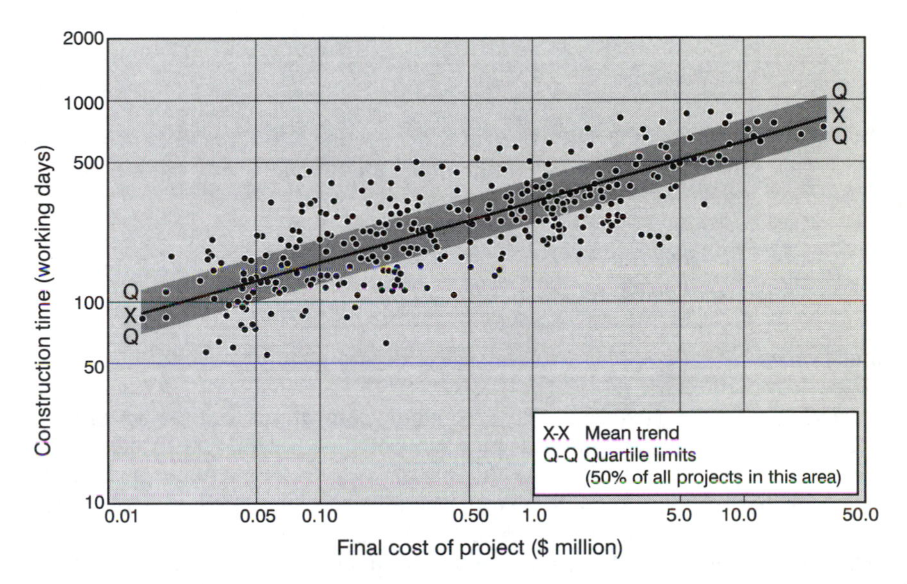

Figure 5.1 Bromilow's original data set, showing the spread of the transformed data and the quartile limits (Bromilow and Henderson, 1977)

The Bromilow equation has the advantage that it is of natural logarithmic form, and that if the data is transformed to the natural logarithm, a straight line relationship should result. This transformed data set may therefore be subjected to linear regression in order to identify the parameters. The equation of the linear relationship is as follows:

$$\ln T = \ln K + B \ln C \tag{5.2}$$

This transformed relationship appears linear, as shown in Figure 5.1 and the corresponding non-linear real relationship together with quartile limits is shown in Figure 5.9 (page 155).

It is interesting to reflect on what is being measured when looking at the relationship between time and cost. Bromilow (1969) was seeking a measure of contract time performance, as the average contract time over-run approximately 20%. His concern was that it was not possible to assess whether a contractor had performed poorly without some standards of time performance for comparison.

Bromilow was seeking a relationship between the duration of construction and the size of a project. He recognised that there are many ways to measure the size of the project, but found that no single indicator was perfect, with the best being the final cost which, he argued, reflects complexity and quality as well as physical size.

One indicator which Bromilow proposed was the 'time performance ratio' (*TPR*) which expressed forecast (contract) duration as a ratio of actual (final) construction time.

$$TPR = \frac{\textit{Contract duration}}{\textit{Final duration}} \tag{5.3}$$

This expression produces a ratio where a *TPR* less than 1.0 represents poor performance. This often seems counter-intuitive, and an alternative is to indicate a time performance indicator (*TPI*):

$$TPI = \left(\frac{\textit{Final duration}}{\textit{Contract duration}} - 1\right)100\% \tag{5.4}$$

which is negative for good performance (% under-run) and positive for poor performance (% over-run).

Final cost is the most frequently used single comparator for time performance. It is rare indeed to find an alternative approach, although Ireland (1986) succeeded in developing a relatively simple model utilising physical factors, with his investigation of building performance for US versus Australian projects for a single company (Civil & Civic Pty Ltd). He was concerned with gross floor area per day and number of storeys. He identified a fascinating relationship, using speed rather than duration for time performance, of the following form:

$$Log_{10} SPEED = -5.72956 + 2.96889(Log_{10} AREA)^{0.6124} + \frac{2.93390}{Storeys} \tag{5.5}$$

It is disappointing that this promising model has not been further explored. The concept of rate of production (speed) which is implicit in all the models, is more directly expressed in this form. It would be worthwhile extrapolating this to examine the change in the rate of construction during the life of the project, which would follow the underlying periodic bell–curve distribution implicit in the cumulative cash flow S curve.

Multi-variate relationships

Researchers have also examined multi-variable influences on project duration in an attempt to explain and remove the spread of projects about the trend. The underlying question here is what factors influence the duration—and thus might be considered in a more sophisticated model. There is no consensus in the research, but the types of factors are discussed here to illustrate the breadth of contributing issues which have been considered. Ireland (1983) cross-correlated eighteen management factors with construction time (refer Table 5.1). Kumaraswamy and Chan (1995) illustrated their view of the influencing factors in a breakdown structure (Figure 5.2). Yeong's proposed model for contract performance is illustrated in Figure 5.3. These illustrate clearly the complexity of factors which might be considered in modeling time performance, and go a long way toward explaining the wide variance from the model which may be observed between the fitted model and the source data[1]. Most interestingly, these are generally factors which are ignored in both time and cost models.

Table 5.1 Interacting managerial factors affecting project duration (Ireland, 1983)

Managerial factor	Interval	Ordinal
Architectural quality		O
Building cost	I	
Building cost per square metre	I	O
Complexity of form of construction		O
Construction coordination		O
Construction planning during design		O
Construction time	I	
Construction time per square metre	I	
Contract variations per unit of building cost	I	
Design coordination		O
Design–construction interface coordination	I	O
Extension of time through industrial disputes	I	
Generation of alternative designs		O
Gross area	I	
Income per square metre	I	
Number of storeys	I	
Quality control on site		O
Use of nominated subcontractors	I	

For the purpose of indicating project time, researchers have not yet been able to determine a better single measure of project size than project cost.

The final cost referred to by Bromilow is the final cost to the client of the building work. To be strictly correct, this is the final price rather than cost, accepting that cost is normally the contractor's outgoings, and that price is the marked-up charge to the client. Kaka and Price (1991) referred to final project 'value' interchangeably with cost. The concept of value includes all sorts of qualitative issues, which is over complex for what is in essence a simple extraction of a one-line budget item. It seems pedantic to try to change the conventions of the research community, and for simplicity, the final project cost will be used here.

1 There are many studies of exploring the factors affecting time performance. Further information can be found in Sidwell (1981), Walker (1995), and Yeong (1994).

- CONSTRUCTION PROJECT DURATION
  - Construction COST/VALUE
  - Type of construction
    - Product, e.g. earth dam; steel framed building
    - Technical parameters, e.g. height; floor area; spans
    - Quality
      - of construction required
      - of design and documentation
      - Others
    - Complexity (scale 1 to 10)
    - Others
  - Location
  - Client's and other imperatives/priorities
  - Total factor productivity
    - Managerial
      - Abilities
      - Motivation
      - Systems
      - Others
    - Organisational
      - Structure
      - Style
      - Information systems
      - Others
    - Labour
      - Work systems
      - Skills
      - Motivation
      - Others
    - Technology
      - Labour/Equipment mix
      - Plant and equipment
        - Age
        - Level of technology
        - Others
      - Others
    - Others
  - Others
  - Type of Contract
    - Risk allocation, e.g. inflation; technical
    - Tenderer selection method (Open, Prequalification, selection, etc.)
    - Management structure, e.g. Traditional; Design and build
    - Payment modalities, e.g. Fixed price; Cost plus; BOT
    - Others
  - Post-contractual developments
    - Variation orders
      - Magnitude
      - Interference level
      - timing
      - Others
    - Conflicts
      - Others
    - Others

Figure 5.2 Factors affecting construction project duration
(adapted from Kumaraswamy and Chan, 1995)

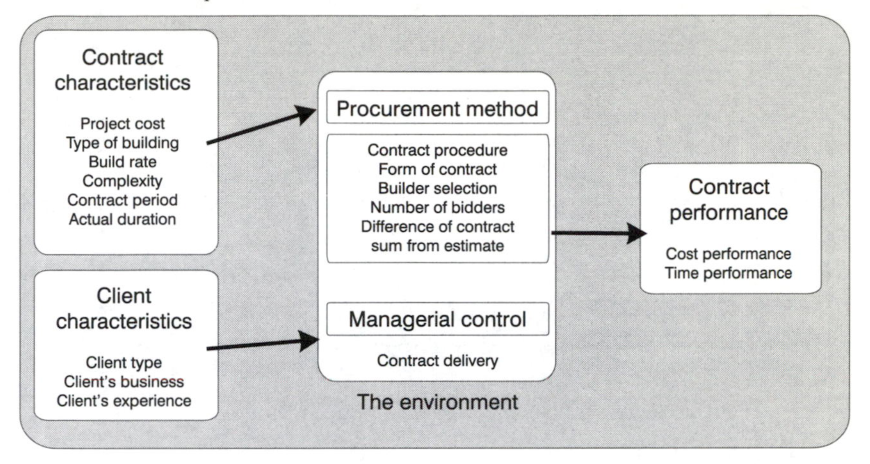

Figure 5.3 Yeong's model for contract performance (adapted from Yeong, 1994)

WORKED CALCULATION OF *B* AND *K*

To demonstrate the calculation of *B* and *K* for the relationship between time and cost given in Equation (5.1), data was sourced from eleven Australian public sector projects, consisting of six police stations, two schools and three institutional buildings. The estimates and actual final values for the data is summarised in Table 5.3 with the adjustment to a common base index (January 1998 = 807). The formula for the index adjustment is:

$$Adjusted\ final\ cost = final\ cost\left(1+\frac{(I_1 - I_2)}{I_2}\right) \tag{5.6}$$

where I_1 is the Index at the date of the project and I_2 is the comparative value of the index.

Table 5.2 calculates from the adjusted final cost the parameters of the time cost relationship for the small data set from Table 5.3. This is the 'new data set' included in Tables 5.4 and 5.5. Table 5.2 demonstrates the transformation of the data into the natural logarithms ($\log_e(C)$ and $\log_e(T)$) which are the transformed (linear) data, from which are calculated the slope (*B*) and the intercept (*Int*). The slope is the natural logarithm of *K* and from this *K* is calculated by Equation 5.7:

$$K = \mathrm{e}^{Int} \tag{5.7}$$

The solid line in Figure 5.4 represents the model derived from the analysis of the data, which has been calculated from the forecast project duration in increments of $0.5 million. The points represent the originating data. This visually illustrates the quality of fit for the model against the data.

Table 5.2 Calculation of B and K from a sample of 11 government sector projects

Project	Index date	BCI	Estimated cost	Estimated days	Final cost	Actual days	TPI	Log_e (Cost)	Log_e (Time)
1	Mar 07	800	$349,809	100	$360,202	100	0%	(1.021)	4.605
2	May 97	802	$1,187,396	150	$1,264,010	173	15%	0.234	5.153
3	Jul 97	804	$3,364,854	272	$3,354,416	280	3%	1.210	5.635
4	Apr 97	800	$407,357	120	$441,000	152	27%	(0.819)	5.024
5	Nov 97	805	$884,253	130	$884,525	153	18%	(0.123)	5.030
6	Mar 97	800	$438,335	100	$438,132	159	59%	(0.825)	5.069
7	Apr 97	800	$1,805,403	170	$1,886,153	193	14%	(0.635)	5.263
8	Sep 97	804	$438,583	150	$538,385	190	27%	(0.619)	5.247
9	Sep 96	785	$17,704,205	263	$21,789,343	531	102%	3.081	6.275
10	Jan 98	807	$1,963,986	143	$2,044,000	153	7%	0.715	5.030
11	Apr 97	800	$4,556,088	222	$4,719,311	303	36%	1.552	5.714
						Average	28%	Slope (***B***)	0.32
								Intercept (***Int***)	5.16
								K= $_e$*Int* =	174

Table 5.3 Adjustment for Building Cost Index (BCI) for the 11 government sector projects

Project	Index date	BCI	Estimated cost	Adjusted Estimated cost	Final cost	Adjusted Final cost
1	Mar 07	800	$352,870	$349,809	$363,354	$360,202
2	May 97	802	$1,194,799	$1,187,396	$1,271,890	$1,264,010
3	Jul 97	804	$3,377,409	$3,364,854	$3,366,932	$3,354,416
4	Apr 97	800	$410,921	$407,357	$444,858	$441,000
5	Nov 97	805	$886,450	$884,253	$886,723	$884,525
6	Mar 97	800	$442,170	$438,335	$441,966	$438,132
7	Apr 97	800	$1,821,200	$1,805,403	$1,902,657	$1,886,153
8	Sep 97	804	$440,220	$438,583	$540,394	$538,385
9	Sep 96	785	$18,200,374	$17,704,205	$22,400,000	$21,789,343
10	Jan 98	807	$1,963,686	$1,963,986	$2,044,000	$2,044,000
11	Apr–97	800	$4,595,954	$4,556,088	$4,760,605	$4,719,311

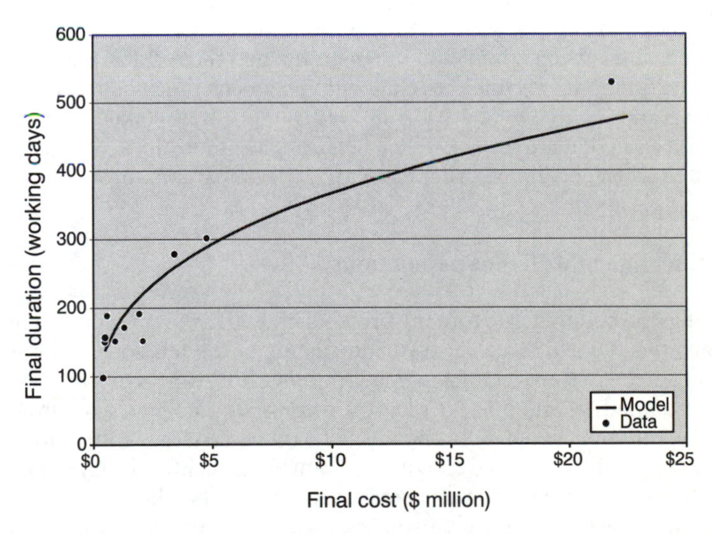

Figure 5.4 Fitted model, with the originating projects for the 11 government sector projects

COMPARATIVE ANALYSIS OF PAST STUDIES

Comparative studies are used in many disciplines to further the knowledge of the discipline. It has long been realised that individual studies, which are most often necessarily small, may not directly lead to significant results, and yet they may contribute collectively to such an outcome. Analysis of collective studies is usually referred to as meta-analysis. It is generally agreed in the literature that Glass (and colleagues) coined the term 'meta-analysis' in 1976.

> Meta-analysis refers to the analysis of analyses…the statistical analysis of a large collection of analysis results from individual studies for the purpose of integrating the findings. It connotes a rigorous alternative to the casual, narrative discussions of research studies which our attempts to make sense of the rapidly expanding research literature. (Glass 1976)

Since that time, meta-analysis[2] has become a widely accepted research tool. A search of the Education Resources Information Centre (ERIC) on-line database (Bangert-Drowns and Rudner, 1991) identified over 800 articles written after 1980 that used or discussed meta-analysis. Lyons (1998) illustrated the growth in the use of the technique in the fifteen years to 1990 and noted the growth was almost geometric.

It is now possible to undertake a meta-analysis of studies conducted over the past 30 years, and comment on the trends in the time–performance parameter. To facilitate the analysis of long-term trends in time–performance and to enable forecasting, a revision to the commonly accepted model for the time–cost relationship is developed. The revised model removes the effects of the price–index from the model by adjusting the time–performance parameter for the interaction between the effects of the index and the sensitivity of time performance to price level.

The adjusted indicator for time–price performance is a better identifier of a long-term trend in the time–performance of building contracts. The bulk of work in this field to date has been completed in Australia, and the results of a re-analysis of available Australian studies into the relationship between time and price is presented. This analysis more clearly shows a small long-term improvement in productivity. This analysis will compare the Bromilow performance indicators for past studies and by adjusting for the cost index bring them all to a comparable form.

The identification of different populations

There is a general belief in categorising projects into different project type or industry sector groupings to yield more accurate forecasting of the relationship between time and cost. The interest in factors which influence the time–cost relationship has frequently resulted in attempts to identify groups or categories of project which behave differently, for which the relationship parameters are significantly different; groups which may be described as coming from significantly different populations. The underlying epistemology here is nomothetic, that is breaking down the data into more and more detail will provide sufficient detail to explain variance while still studying averages of categories of projects. The nomothetic epistemology is a relevant approach to essentially a single point relationship, although the use of

[2] For a comprehensive discussion of the methodology of meta-analysis and its application to the construction industry research, refer to Horman (2000). A proposal for increased meta-analysis in construction research may be found in Kenley (1998).

characterisation of projects is uncertain, not yet having been proven in the literature by demonstrating any such characterisations are statistically significant (although, Ng *et al.* (2001) argued that exclusion of industrial projects resulted in significant differences from when they were included).

There is not yet any acceptable proof that identified differences are anything more than differences in sampling from a single population. Sampling error is the most likely reason for differences between different sets of results, whether for different categories or for different years. However, this does not mean that there are not valid reasons to examine differences. Bromilow (1969) argued:

> in many of the projects represented in the lower quartile [better performance] in Figure 1 [reproduced here as Figure 5.1] the builder has been directly nominated, and others are schools or offices (but not banks) under restricted tender. There is a possibility that such projects may have a different performance pattern...

Previous categorisations have differentiated fast, average and slow projects (Bromilow and Henderson, 1977); steel frame from concrete frame (Ireland, 1986); the frequently used public and private sector projects (Bromilow, *et al.,* 1980, 1988; Kaka and Price, 1991; Kumaraswamy and Chan, 1995; Chan, 1999; Yeong, 1994; Ng *et al*., 2001); the economic cycle (Mak *et al.*, 2000) and industrial and non-industrial projects (Ng *et al.*, 2001). These characterisations have been inconclusive but there has been a wide acceptance of the differentiation between public and private sector projects. The series of investigations conducted over 20 years by Bromilow and the Australian Institute of Quantity Surveyors have had relatively large sample sizes. It seems from their data that such a characterisation is likely to be valid. It is also the only characterisation which has enough studies to allow the identification of long-term trends. For these reasons, differentiation by public and private sector will be used in the following sections.

Impact of cost escalation

Cost or price escalation affects the interpretation of time–cost models. This is a confusing issue not well handled in the literature. This section will discuss the impact of inflation on cost models, will introduce cost or price indices and show how the Bromilow time–cost model is adjusted for changes in the index for comparative purposes.

Over time, the buying power of one dollar for building work varies. Generally, but not always, inflation works to reduce the amount of building work which may be purchased with one dollar. The concept is the same as a typical consumer price index used by economists to understand changes in the economy, but there are a number of key differences. If a building was constructed entirely of components requiring no on-site construction, then the normal consumer price index model of a 'basket of goods' might work for modelling construction costs. However, this is far from the case, and changes in the labour market and employee productivity are major factors in a building index. Also, the building market is a highly specialised industry which suffers severely from fluctuations in supply and demand, which can exaggerate or even counter inflationary effects.

Building cost indices are referred to as composite cost indices as they are commonly made up of a mix of costs relevant to site construction: materials; plant; and labour. They are designed for adjustment of contract prices during the life of a project, through contract provisions known as 'Rise and Fall'. As such, they are intended to be accurate for adjusting prices relative to the start of the contract. There is much less certainty about their accuracy for measuring long-term trends, due to the uncertain impact of productivity.[3]

One area of confusion with building cost indices is the question as to whether the cost indices include an adjustment for productivity change. In an article about their own Building Cost Index, Irwin (1980, p. 84) of the AIQS declared that:

> indices which are quite suitable for calculation of rise and fall on individual contracts are not relevant in the long-term because of the effect of increased productivity, changes in regulations and varying market conditions.

Under a broad heading, productivity included:

> efficiencies obtained through improved management practices, more efficient equipment (from caulking guns to hoists and concrete pumps), new materials, a greater appreciation of designers on the need to standardise, more sophisticated prefabrication techniques, etc.

Ferry (1970) makes it quite clear that productivity is a major problem, as there is no clear method for its inclusion in the design and use of building cost indices. This is a major issue when considering whether or not productivity changes are revealed. This issue will be revisited in the meta-analysis of time–cost studies. First, however, it is necessary to examine the use of building cost indices to remove the effects of price escalation on the time–cost models.

Price escalation indicates that a project working at a later time will have a lower K value. Thus the K value must be adjusted to K' and this adjustment is achieved by the use of a building cost index. Bromilow's equation was:

$$T = KC^{B} \qquad \text{(5.1 repeated)}$$

and as the term which is indexed is C not K, then the recalculated value K' is:

$$K' = K_1 \left(\frac{i_1}{i_2} \right)^{B} \qquad (5.8)$$

where: K_1 is the calculated K value, i_1 is the earlier cost index and i_2 is the later cost index, and where B is the calculated or provided (for past studies) constant. Equation 5.8 is from Kenley (2001) but the use of the terms 'earlier' and 'later' can

[3] Building cost indices are well addressed in the literature. Ferry (1970) Ch. 14 contains an excellent summary of the justification, purpose and make-up of such indices, and Bathurst and Butler (1980) Ch. 15 provide an good discussion of the use of indices and the weighting of the components.

lead to confusion. It is better to determine the base is to be used (this was the later cost index in Kenley, 2001) and express the relationship accordingly.

$$K' = K_1 \left(\frac{i}{i_b} \right)^B \quad (5.9)$$

where i is the index at the time of sampling and i_b at the time of comparison.

Bromilow's work initiated a series of replication projects over 30 years, undertaken within Australia and internationally. No mention was made in Bromilow's 1969 work of the impact of cost escalation on his model. However, it is clear that a model that is based on a constant K, which equates to the time taken to build \$1 million of building work, must be influenced by the amount of work which \$1 million can buy at any given time. If less work is required to complete \$1 million, due to inflation, then it will take less time and K will appear to reduce. In other words, cost escalation erodes buying power, so that without adjustment, the value of K will reduce over time under inflationary conditions. Recognising this, Bromilow and Henderson (1977), and indeed most subsequent authors, developed the practice of restating past K values in adjusted current dollar terms for comparison. Thus we find that the values of $K = 350$ and $B = 0.30$, which were provided without reference to a cost index in 1969 (Bromilow, 1969), were restated into equivalent terms at a September 1972 base by Bromilow and Henderson (1977) as $K = 313$, $B = 0.30$. This reflected that \$1 million was able to buy less in 1972 than it had in 1969, accordingly it took less time to build and the K value was lower. Interestingly, it is not possible to replicate these figures unless one calculates the original base used by Bromilow as June 1965. In their later report, Bromilow, *et al.* (1988) showed that the original $K = 350$ was reduced to $K = 204$ when adjusted to January 1986 prices.

Bromilow and Henderson (1977), Bromilow *et al.* (1980), Ireland (1983) and Yeong (1994) proposed the use of the Australian Building Economist Cost Index (BCI) produced by the Australian Institute of Quantity Surveyors (and which was dominated by Melbourne data in its early days). Bromilow *et al.* (1988) used the Construction Price Index from the Australian Department of Housing and Construction. Ng *et al.* (2001) proposed the commercially available *Rawlinson* price index (RPI), presumably the set for Sydney[4]. A calculation of the original Bromilow study of data from 1964 to 1967 recalculated to September 1972 is marginally more accurate using the RPI than the BCI. However, subsequent re-calculations of the September 1972 values to later values are more accurate using the BCI. The BCI will be used here.

Using Equation 5.8 or 5.9, it is now possible to look at past analyses and to compare them in equivalent terms. In response to a paper attempting just such a long-term analysis Ng *et al.* (2001), Kenley (2001) conduct a detailed analysis of all Australian studies to derive K' with all results brought to a common index at BCI = 807 (January 1998). The results for: all projects; government sector projects; and private sector projects are tabulated in Table 5.4. This tables includes the small

4 This index is replicated in many countries and may prove suitable for subsequent international analysis.

sample of projects calculated earlier in this chapter. The corresponding results for each grouping are displayed in Figures 5.5, 5.6 and 5.7 respectively, plotted against the mid date of the data set. This illustrates the long-term trend in the performance parameter K over 30 years.

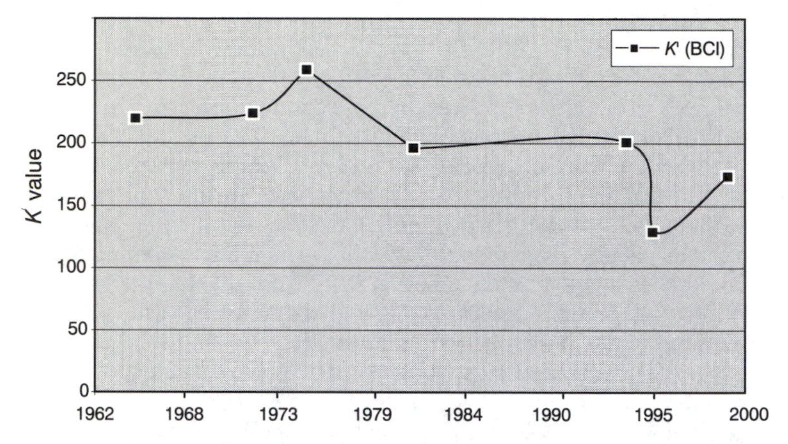

Figure 5.5 K' for all Australian public projects as at January 1998 (adapted from Kenley, 2001)

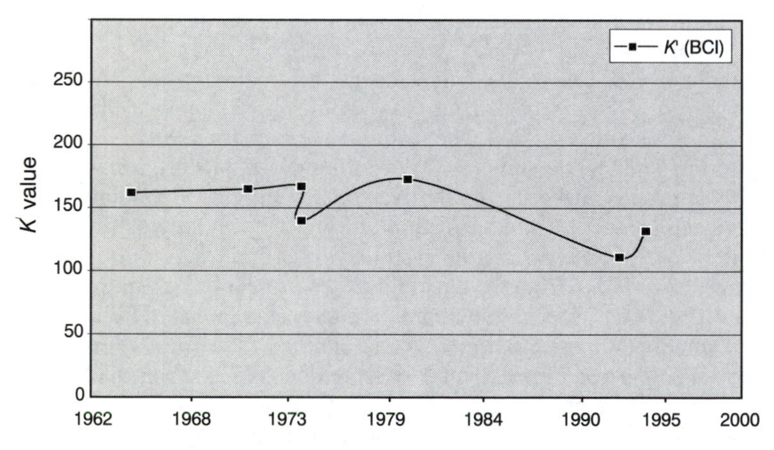

Figure 5.6 K' for all Australian public projects as at January 1998 (adapted from Kenley, 2001)

Table 5.4 shows the corresponding values of K' for BCI at base BCI = 100. This is useful as a universal base and allows calculation of a forecast at any time using only the new BCI figure.

It is important, when looking for trends involving indexed data, to be careful about dates. Table 5.4 includes the dates of the sample data in the various studies, as well as the date to which these studies were indexed by their authors. In some cases the discrepancy between the two may be misleading.

Table 5.4 Comparison of adjusted K' at January 1998

Source	Data set date range	Mid date	BCI date	BCI	Sample size			Parameter B			Parameter K'			Note
					All	Gov	Pvt	All	Gov	Pvt	All	Gov	Pvt	
New Data set	1997–1998	Jan 98	Jan 98	807	11	11		0.33	0.33		172	172		
Ng *et al.* (2001)	1991–1998	Jun 94	Mar 98	807	93	31	62	0.31	0.32	0.30	130	128	131	
Yeong (1994)	1990–1993	Dec 92	Dec 92	737	87	67	20	0.22	0.24	0.37	187	199	110	1
Bromilow *et al.* (1988)	1976–1986	Jan 81	Jan 81	584	777	683	94	0.37	0.38	0.28	171	194	191	2
Ireland (1983)	1970s	Jan 75	Jan 75	312	25		25	0.47		0.47	138		138	3
Bromilow *et al.* (1980)	1974–1976	Jan 75	Jan 75	210	160	74	86	0.36	0.34	0.37	179	256	165	2
Bromilow *et al.* (1980)	1970–1973	Jan 72	Jan 72	210	235	203	32	0.28	0.28	0.28	192	222	164	2
Bromilow (1969)	1964–1967	Jun 65	Jun 65	110	328	187	141	0.30	0.30	0.30	184	218	161	4
Weighted Average					1716	1256	460	0.33	0.34	0.32	175	204	161	

1. Using Yeong's 5/7 adjustment for working days
2. Calculated from the components using Bromilow *et al.* (1980) method of weighting
3. It is likely that Ireland's data sample was from the private sector
4. Breakdown published in Bromilow *et al.* (1980)

The size of the sample is also important. It is clear that the various Bromilow surveys had much larger samples than subsequent surveys, particularly in the public buildings sector. It is possible that some of these subsequent result sets are suspect, due to the small number of projects and the wide time span involved (certainly the eleven projects added in this chapter should not be relied upon, but should be indicative only). For example, Ireland's data was collected during the 1970s with two of the sample of 25 being completed prior to 1970. This is an enormous range and includes data sets of only 2.5 projects average per year during a period of large economic change and high inflation. The importance of this particular data set should therefore be discounted in interpreting a trend.

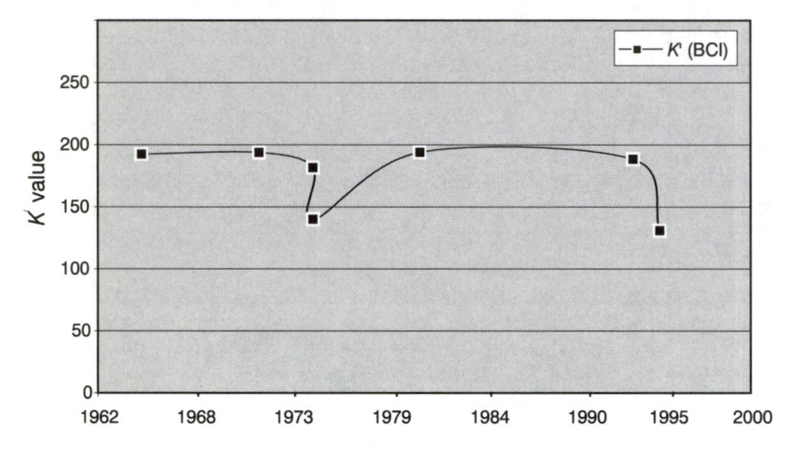

Figure 5.7 K' for all Australian projects as at January 1998 (adapted from Kenley, 2001)

This result is inconclusive about the possibility of a real long-term trend in productivity for either of the two sectors, public and private. Only two data sets are significantly lower than the weighted average of K values (weighted by the size of the sample). In fact, the weighted average for K' is almost the same as the original result obtained by Bromilow (1969). There also appear to be outliers in the chart, however, these two data sets (Ireland and Yeong) have B values well away from the normal 0.3 This is discussed further in the following section.

There are three conclusions which may be drawn from these results. The first is that there has been no real improvement in productivity identified over the past 30 years. This means that in real terms, it takes just as long to spend money in construction as it did 30 years ago and that the Australian industry has not improved significantly.

A second interpretation is that any productivity gain which has been obtained, is in fact already included in the BCI.

A third view is that there is a problem with the model, due to the differing B values in two data sets. This latter point can be tested to some extent by calculating to the base of the index as in the following section.

A revised model

It has been shown here how the K values may be adjusted and updated to reflect changes in a building cost index. Bromilow's original work claimed that a constant had been found which was a 'constant characteristic of building time performance in Australia' (Bromilow and Henderson, 1977). The confusion of having a constant which not only varies, but which has to be recalculated, is most likely the reason for limited understanding and acceptance of this otherwise consistent model. Consequently, a new model is sought which calculates a true 'constant' characteristic of building performance.

The reason that K is a variable constant is because it is subject to the influence of the building cost index. Therefore, it would be better to re-express the Bromilow model with the cost index directly incorporated, so that a true constant for building performance is expressed. This constant, equivalent to K', would be the K value if it were expressed at the base of the index (BCI=100) as in Table 5.5. Here this constant is expressed as K_2, where K_2 is K' in the special case where BCI_2 =100. First K_2 is calculated as follows:

$$K' = K_1 \left(\frac{i}{i_b} \right)^B \qquad \text{(5.9 repeated)}$$

$$\Rightarrow K_2 = K_1 \left(\frac{BCI_2}{100} \right)^B \qquad (5.10)$$

Table 5.5 Comparison of adjusted K' at base 807 with base 100

	K' base 807			K_2 base 100		
Source	All	Gov	Pvt	All	Gov	Pvt
New Data set	172	172		343	343	
Ng *et al.* (2001)	130	128	131	250	252	247
Yeong (1994)	187	199	110	295	329	239
Bromilow *et al.* (1988)	171	194	191	417	434	310
Ireland (1983)	138		138	374		374
Bromilow *et al.* (1980)	179	256	165	382	527	362
Bromilow *et al.* (1980)	192	222	164	348	403	297
Bromilow (1969)	184	218	161	360	412	304
Weighted average	175	204	161	377	420	309

This allows a universal constant to be used for forecasting the duration of a project from its project value. The results are presented in Table 5.5.

From this table, using the weighted average mean K_2 values and where BCI is the current BCI, C is current cost and B is 0.34, 0.32 and 0.33 and K_2 is 420, 309 and 377 for government, private and all sector projects respectively.

The K_2 values exhibit slightly greater downward trend (improved production efficiency) than K' from Figure 5.4. These are illustrated in Figure 5.8 for the three sample sets and, interestingly, the apparent outliers have reduced in significance. This suggests that B has an important effect on the long-term adjustment. Without these two data sets, the trend effectively disappears as before.

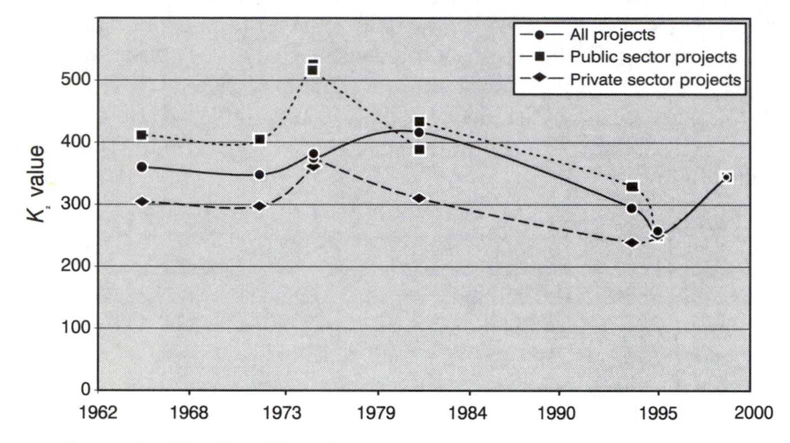

Figure 5.8 K' for all, public and private Australian projects with index base 100 (adapted from Kenley, 2001)

Further work is required to understand the operation of the exponent B, a constant that describes how the time performance was affected by project cost. B represents the slope of the linear equation for the transformed data and the rate of curve of the raw data. These results suggest that further investigation would be valuable into why some data sets (usually small ones) derive a different B value—with corresponding change in K.

Variance in the model

There is a large amount of variation inherent in the model. Bromilow and Henderson (1977) referred to fast and slow projects ($K = 250$, $K = 407$ at September 1972 while $B = 0.3$, which equates to $K' = 278$, $K' = 452$ respectively). While not valid as a project characterisation, nevertheless this division is very informative about the model and the accuracy of fit to real projects.

Figure 5.1 illustrated the Bromilow data set on a logarithmic scale. The quartile limits equate to the fast and slow projects. What is significant here is that 25% of projects can be expected to be faster than the 'fast' project and 25% can be expected to be slower than the 'slow project'. This is a very large variation possible from the forecast. Consider Figure 5.9, from which it can be calculated that in 1977 a project with a cost of $8 million would have had a mean duration of 650 days. However, only 50% of projects would have been within the range of 500 days (expected for a project of only $3.5 million) to 820 days (expected for a project of approx $15 million). The remaining half could be expected to be outside that range. It must be recognised therefore that this is a model with a high degree of uncertainty.

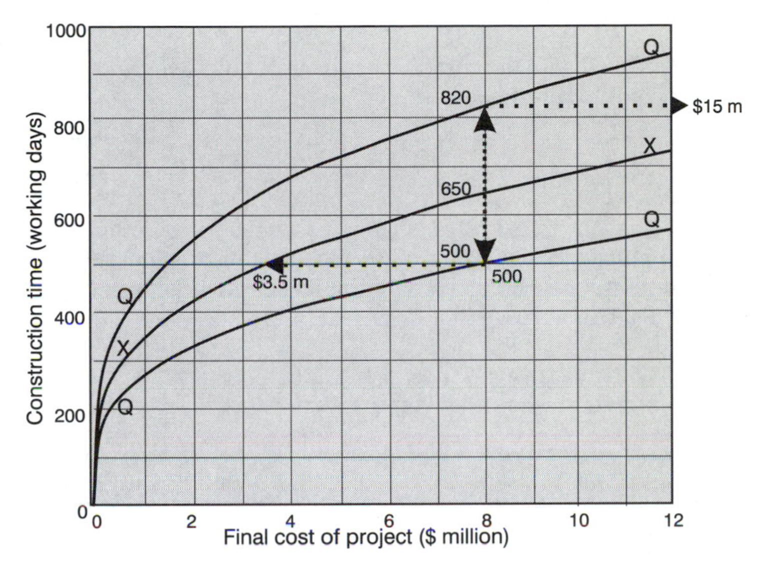

Figure 5.9 Quartile limits expressed on a normal scale, with sample 50% range marked. (adapted from Bromilow, 1969)

International comparisons

Several studies have replicated the Bromilow study internationally, however:

> direct comparison of *K* values is unrealistic considering disparities in currency values and construction cost levels between different economies (Kumaraswamy and Chan, 1995).

This statement applies equally to K'. Despite the difficulty, it remains worthwhile to explore the significance of an international comparison. Kumaraswamy and Chan also ask 'is the speed of construction really remarkable in Hong Kong'? This very interesting question can be repeated in any region, so a method for comparative analysis is desirable. To do this, a further adjustment is made to the model, to re-express the results in an equivalent currency. This is required for the same reason as the cost index adjustment, the efficiency indicator *K* is dependent on the speed of construction of a fixed value of construction, typically $1 million at the time of the base index. Ideally this analysis requires an internationally reproducable price index. In the absense of this, all identified *K* values have been converted to Australian dollars using the exchange rate applicable at the time (refer to Table 5.7, page 158), and then adjusted to the common index BCI=base 100.

Returning to the Bromilow equation and the adjustment for the price index, it can be seen that a similar adjustment is required for the currency exchange. As the term which is indexed is *C* not *K*, then the recalculation to include currency exchange *J* is:

$$J = K_1 (r_x)^B \tag{5.11}$$

If this is then combined with the cost index adjustment, for convenience using base 100, then the composite adjustment is:

$$J = K_1 \left(\frac{BCI.r_x}{100} \right)^B \tag{5.12}$$

Table 5.6 summarises the results of the available international studies, with *B* and the original *K* in local currencies, and the calculated *J* adjusted to Australian dollars and with the BCI at base=100.

Table 5.6 Comparison of adjusted K' at Jan 1998

	B			*K*			*J*		
Source	All	Gov	Pvt	All	Gov	Pvt	All	Gov	Pvt
Australia									
Aggregated results	0.33	0.34	0.32	253	248	199	377	420	309
Hong Kong									
Chan (1999)	0.29	0.28	0.34	152	166	120	427	444	306
Malaysia									
Yeong (1994)	0.352	0.352		518	518		1046	1046	
Chan (2001)	0.32	0.32		269	269		649	649	
UK									
NEDO (1988)			0.247			305			392
Kaka and Price (1991)—Fixed		0.318	0.212		399	274		597	359
Kaka and Price (1991)—Indexed		0.205	0.082		487	491		632	545

Note: Yeong (1994) and Chan (2001) is the same data set. Yeong (who analysed the data for a Master of Applied Science) converted the data to Australian dollars. Thus the great difference in the *K* values between Yeong and Chan. However, the data is not presented and it is thus not possible to further reconcile these results which yield an inconsistent value of *J*.

While there are not enough results here for a clear trend to emerge, J does appear to be a successful indicator of comparitive international performance. Consistent results are achieved, except Malaysia which is clearly higher. At issue here, however, is the difference in the cost of labour which would distort the results. This supports the need for a true comparison between countries based on comparative costs.

This may in part answer the question posed by Kumaraswamy and Chan (1995), who asked whether 'the speed of construction was really remarkable [faster] in Hong Kong'? While they did not attempt to answer the question because of the complexity of international comparisons, the results in this analysis suggest that after currency exchange and cost indexing, Hong Kong is no different from Australia or the UK. This conclusion does not take into account the possibility that there may be differential in purchasing power between these countries, and therefore the results are indicative only. Further research is required into this relationship.

ALTERNATIVE MODELS

Berny and Howes (1982) followed Hudson (1978) in being able to derive a forecast end-date based on in-project cash flow data (refer to Chapter 4). They each used the standard curve profiles for a particular value project (refer to Chapter 3) as the basis for this calculation. Implicit in this approach, is that with a new revised project end date, there is a new S-curve or cash flow profile.

Alternatively, Hudson noted that when a project has been underway for a few months, it is possible to determine the C and K parameters for that project. This is done by solving the simultaneous equations formed by substituting two actual values into the cash flow formula. Hudson used the two most recent values, but to accommodate irregularities proposed smoothing the data points with polynomial regression.

This complex procedure was not followed by Berny and Howes, who did not publish a method for revising the cash flow profile, although a similar method may have been incorporated into their cash flow software.

CONCLUSION

Bromilow's time–cost model has stood the test of time and has proved to be a remarkable contribution to our understanding of construction. It has spawned a series of research projects, providing information to clients and contractors around the world, and been shown to be the best indicator of project duration from cost available.

While there are many factors which influence project duration, it seems that they are all governed by this simple relationship. It is exciting that attention has turned, in recent times, to using this relationship to explore matters of long-term productivity improvement and the differences in productivity between countries. Productivity research has its own methods and devotees, but this independent method may prove significant in measuring change at a global level.

This chapter concludes the sections on individual projects, by concluding the discussion of the relationship between gross cash flow and time—which is a time–cost relationship.

Table 5.7 Exchange rates used for international comparisons
Australian dollar to British pound,
Hong Kong dollar and Malaysian ringgit.

Month	AUD/ GBP	AUD/ HKD	AUD/ MYR	Month	AUD/ GBP	AUD/ HKD	AUD/ MYR	Month	AUD/ GBP	AUD/ HKD	AUD/ MYR
Jan 81	0.4919	6.1252	0.3807	Apr 87	0.4361	5.5506	0.5655	Jul 93	0.4533	5.2575	0.5746
Feb 81	0.5068	6.1632	0.3801	May 87	0.4285	5.5743	0.5655	Aug 93	0.4542	5.2509	0.5786
Mar 81	0.5210	6.1648	0.3769	Jun 87	0.4408	5.6054	0.5554	Sep 93	0.4274	5.0431	0.6023
Apr 81	0.5301	6.1783	0.3744	Jul 87	0.4400	5.5280	0.5558	Oct 93	0.4400	5.1100	0.5938
May 81	0.5462	6.2132	0.3748	Aug 87	0.4421	5.5226	0.5576	Nov 93	0.4489	5.1363	0.5889
Jun 81	0.5779	6.2906	0.3745	Sep 87	0.4419	5.6716	0.5462	Dec 93	0.4517	5.2032	0.5768
Jul 81	0.6099	6.5175	0.3721	Oct 87	0.4279	5.5528	0.5556	Jan 94	0.4665	5.3774	0.5289
Aug 81	0.6262	6.7692	0.3695	Nov 87	0.3864	5.3486	0.5833	Feb 94	0.4841	5.5392	0.5055
Sep 81	0.6330	6.9215	0.3702	Dec 87	0.3886	5.5232	0.5642	Mar 94	0.4765	5.4930	0.5177
Oct 81	0.6211	6.8442	0.3805	Jan 88	0.3949	5.5375	0.5537	Apr 94	0.4828	5.5294	0.5197
Nov 81	0.6021	6.4928	0.3869	Feb 88	0.4061	5.5676	0.5426	May 94	0.4815	5.5961	0.5276
Dec 81	0.5958	6.3871	0.3923	Mar 88	0.3998	5.7187	0.5311	Jun 94	0.4802	5.6660	0.5260
Jan 82	0.5907	6.4572	0.3976	Apr 88	0.3983	5.8432	0.5193	Jul 94	0.4746	5.6720	0.5250
Feb 82	0.5874	6.3860	0.3989	May 88	0.4158	6.0758	0.4977	Aug 94	0.4799	5.7189	0.5271
Mar 82	0.5873	6.1813	0.4054	Jun 88	0.4545	6.3052	0.4788	Sep 94	0.4738	5.7338	0.5270
Apr 82	0.5934	6.1271	0.4065	Jul 88	0.4692	6.2508	0.4759	Oct 94	0.4594	5.7022	0.5296
May 82	0.5852	6.0967	0.4121	Aug 88	0.4749	6.2885	0.4680	Nov 94	0.4750	5.8358	0.5174
Jun 82	0.5878	6.0564	0.4141	Sep 88	0.4700	6.1821	0.4742	Dec 94	0.4965	5.9884	0.5042
Jul 82	0.5825	5.9668	0.4200	Oct 88	0.4656	6.3256	0.4611	Jan 95	0.4856	5.9218	0.5117
Aug 82	0.5672	5.9289	0.4344	Nov 88	0.4704	6.6435	0.4390	Feb 95	0.4737	5.7576	0.5261
Sep 82	0.5597	5.8693	0.4420	Dec 88	0.4695	6.6923	0.4331	Mar 95	0.4590	5.6790	0.5347
Oct 82	0.5562	6.2307	0.4474	Jan 89	0.4908	6.7940	0.4220	Apr 95	0.4577	5.6888	0.5484
Nov 82	0.5776	6.2901	0.4486	Feb 89	0.4884	6.6807	0.4276	May 95	0.4581	5.6250	0.5571
Dec 82	0.5991	6.3337	0.4390	Mar 89	0.4768	6.3693	0.4446	Jun 95	0.4512	5.5665	0.5696
Jan 83	0.6236	6.4117	0.4459	Apr 89	0.4724	6.2535	0.4574	Jul 95	0.4563	5.6329	0.5607
Feb 83	0.6303	6.3827	0.4548	May 89	0.4744	6.0186	0.4793	Aug 95	0.4732	5.7396	0.5436
Mar 83	0.5932	5.8811	0.4941	Jun 89	0.4869	5.8926	0.4883	Sep 95	0.4835	5.8312	0.5281
Apr 83	0.5648	5.8882	0.4998	Jul 89	0.4651	5.9045	0.4930	Oct 95	0.4798	5.8529	0.5216
May 83	0.5588	6.1202	0.4947	Aug 89	0.4788	5.9613	0.4883	Nov 95	0.4770	5.7640	0.5285
Jun 83	0.5667	6.3879	0.4904	Sep 89	0.4917	6.0331	0.4797	Dec 95	0.4807	5.7274	0.5317
Jul 83	0.5732	6.2747	0.4899	Oct 89	0.4877	6.0450	0.4794	Jan 96	0.4852	5.7355	0.5274
Aug 83	0.5852	6.5434	0.4835	Nov 89	0.4979	6.1184	0.4725	Feb 96	0.4919	5.8425	0.5193
Sep 83	0.5924	7.1086	0.4792	Dec 89	0.4923	6.1380	0.4707	Mar 96	0.5051	5.9649	0.5100
Oct 83	0.6104	7.3962	0.4667	Jan 90	0.4730	6.1016	0.4734	Apr 96	0.5183	6.0770	0.5068
Nov 83	0.6203	7.1551	0.4656	Feb 90	0.4477	5.9304	0.4872	May 96	0.5260	6.1658	0.5032
Dec 83	0.6280	7.0272	0.4745	Mar 90	0.4651	5.9034	0.4871	Jun 96	0.5132	6.1242	0.5062
Jan 84	0.6436	7.0639	0.4715	Apr 90	0.4665	5.9543	0.4803	Jul 96	0.5085	6.1106	0.5082
Feb 84	0.6484	7.2805	0.4579	May 90	0.4537	5.9272	0.4862	Aug 96	0.5052	6.0561	0.5122
Mar 84	0.6535	7.4147	0.4584	Jun 90	0.4555	6.0649	0.4736	Sep 96	0.5084	6.1306	0.5044
Apr 84	0.6496	7.2069	0.4730	Jul 90	0.4370	6.1448	0.4675	Oct 96	0.4991	6.1224	0.5037
May 84	0.6522	7.0820	0.4792	Aug 90	0.4253	6.2842	0.4587	Nov 96	0.4793	6.1611	0.4974
Jun 84	0.6410	6.8958	0.4903	Sep 90	0.4390	6.4067	0.4496	Dec 96	0.4788	6.1621	0.4971
Jul 84	0.6320	6.5501	0.5126	Oct 90	0.4115	6.2224	0.4627	Jan 97	0.4689	6.0184	0.5165
Aug 84	0.6452	6.6418	0.5059	Nov 90	0.3935	6.0248	0.4801	Feb 97	0.4723	5.9477	0.5238
Sep 84	0.6613	6.5160	0.5116	Dec 90	0.4007	6.0102	0.4803	Mar 97	0.4893	6.1000	0.5126
Oct 84	0.6858	6.5442	0.4966	Jan 91	0.4028	6.0746	0.4728	Apr 97	0.4779	6.0336	0.5131
Nov 84	0.6930	6.7188	0.4792	Feb 91	0.3989	6.1068	0.4733	May 97	0.4749	6.0017	0.5146
Dec 84	0.7082	6.5761	0.4927	Mar 91	0.4234	6.0077	0.4730	Jun 97	0.4585	5.8409	0.5268
Jan 85	0.7232	6.3667	0.4946	Apr 91	0.4455	6.0753	0.4665	Jul 97	0.4445	5.7471	0.5221
Feb 85	0.6746	5.7530	0.5315	May 91	0.4492	6.0239	0.4684	Aug 97	0.4617	5.7334	0.4896
Mar 85	0.6194	5.4372	0.5575	Jun 91	0.4606	5.8764	0.4733	Sep 97	0.4516	5.5997	0.4571
Apr 85	0.5320	5.1291	0.6094	Jul 91	0.4673	5.9884	0.4651	Oct 97	0.4407	5.5685	0.4214
May 85	0.5422	5.2632	0.5967	Aug 91	0.4645	6.0742	0.4597	Nov 97	0.4117	5.3756	0.4256
Jun 85	0.5193	5.1678	0.6091	Sep 91	0.4597	6.1531	0.4569	Dec 97	0.3988	5.1268	0.3986
Jul 85	0.5066	5.4230	0.5789	Oct 91	0.4599	6.1452	0.4594	Jan 98	0.4016	5.0837	0.3454
Aug 85	0.5108	5.5080	0.5739	Nov 91	0.4420	6.1033	0.4638	Feb 98	0.4110	5.2207	0.3887
Sep 85	0.5055	5.3819	0.5838	Dec 91	0.4221	5.9952	0.4729	Mar 98	0.4029	5.1866	0.3987
Oct 85	0.4942	5.4730	0.5803	Jan 92	0.4133	5.8023	0.4974	Apr 98	0.3901	5.0551	0.4102
Nov 85	0.4705	5.2866	0.6065	Feb 92	0.4229	5.8326	0.5114	May 98	0.3853	4.8912	0.4147
Dec 85	0.4714	5.3169	0.6044	Mar 92	0.4401	5.8763	0.5114	Jun 98	0.3663	4.6839	0.4134
Jan 86	0.4914	5.4657	0.5833	Apr 92	0.4340	5.9013	0.5139	Jul 98	0.3760	4.7884	0.3891
Feb 86	0.4891	5.4575	0.5789	May 92	0.4177	5.8523	0.5245	Aug 98	0.3603	4.5628	0.4040
Mar 86	0.4824	5.5305	0.5569	Jun 92	0.4073	5.8440	0.5255	Sep 98	0.3501	4.5628	0.4463
Apr 86	0.4823	5.6348	0.5325	Jul 92	0.3885	5.7627	0.5369	Oct 98	0.3647	4.7877	0.4259
May 86	0.4781	5.6780	0.5293	Aug 92	0.3730	5.6040	0.5524	Nov 98	0.3822	4.9162	0.4145
Jun 86	0.4567	5.3808	0.5534	Sep 92	0.3913	5.5848	0.5530	Dec 98	0.3700	4.7893	0.4255
Jul 86	0.4174	4.9147	0.6009	Oct 92	0.4325	5.5253	0.5586	Jan 99	0.3831	4.8971	0.4164
Aug 86	0.4120	4.7761	0.6252	Nov 92	0.4518	5.3355	0.5747	Feb 99	0.3932	4.9586	0.4112
Sep 86	0.4233	4.8540	0.6141	Dec 92	0.4447	5.3394	0.5639	Mar 99	0.3891	4.8883	0.4172
Oct 86	0.4475	4.9787	0.5969	Jan 93	0.4392	5.2074	0.5718	Apr 99	0.3990	4.9752	0.4099
Nov 86	0.4527	5.0254	0.5938	Feb 93	0.4744	5.2812	0.5569	May 99	0.4103	5.1388	0.3970
Dec 86	0.4582	5.1395	0.5840	Mar 93	0.4842	5.4736	0.5423	Jun 99	0.4115	5.0912	0.4010
Jan 87	0.4390	5.1351	0.5887	Apr 93	0.4606	5.5003	0.5452	Jul 99	0.4166	5.0923	0.4010
Feb 87	0.4370	5.2049	0.5892	May 93	0.4514	5.3995	0.5578	Aug 99	0.4014	5.0045	0.4082
Mar 87	0.4315	5.3605	0.5768	Jun 93	0.4475	5.2212	0.5766				

REFERENCES

Bangert-Drowns, R. L. and Rudner, L. M. (1991). 'Meta-Analysis in Educational Research'. On-line publication—*The Eric Database.* http://ericae.net/db/edo/ED339748.htm, or http://ericae.net/pare/getvn.asp?v=2&n=8

Bathurst, P. and Butler, D. A. (1980). *Building Cost Control*. London, Heinemann.

Berny, J. and Howes, R. (1982). 'Project management control using real time budgeting and forecasting models'. *Construction Papers* **2**: 19–40.

Bromilow, F. J. (1969). 'Contract time performance expectations and the reality'. *Building Forum* **1**: 70–80.

Bromilow, F. J. (1970). 'The nature and extent of variations to building contracts'. *The Building Economist* (November): 93–118.

Bromilow, F. J. and Henderson, J. A. (1977). *Procedures for Reckoning the Performance of Building Contracts*. Highett, Commonwealth Scientific and Industrial Research Organisation, Division of Building Research.

Bromilow, F. J., Hinds, M. F., and Moody, N.F. (1980). 'AIQS Survey of building contract time performance'. *The Building Economist* (September): 79–82.

Bromilow, F. J., Hinds, M. F., and Moody, N.F. (1988). *The Time and Cost Performance of Building Contracts 1976–1986*. Canberra, Australian Institute of Quantity Surveyors.

Chan, A. P. C. (1999). 'Modelling building durations in Hong Kong'. *Construction Management and Economics* **17**(2): 189–196.

Chan, A. P. C. (2001). 'Time–cost relationship of public sector projects in Malaysia'. *International Journal of Project Management* **19**: 223–229.

Ferry, D. J. (1970). *Cost Planning of Buildings*. London, Granada Publishing.

Glass, G. (1976). 'Primary, secondary and meta-analysis of research'. *Educational Researcher,* **5**: 3–8

Horman, M. J. (2000). Process dynamics: buffer management in building project operations. PhD thesis, Melbourne, University of Melbourne.

Hudson, K. W. (1978). 'DHSS expenditure forecasting method'. *Chartered Surveyor—Building and Quantity Surveying Quarterly* **5**: 42–45.

Ireland, V. B. E. (1983). The role of managerial actions in the cost, time and quality performance of high rise commercial building projects. Unpublished PhD Thesis, Sydney, Faculty of Architecture, University of Sydney.

Ireland, V. B. E. (1986). An investigation of U.S. building performance, Report, NSW Institute of Technology: pp. 28.

Irwin, K. (1980). 'Building cost indices'. *The Building Economist* (September): 83–85.

Kaka, A. P. and Price, A. D. F. (1991). 'Relationship between value and duration of construction projects'. *Construction Management and Economics* **9**(4): 383–400.

Kenley, R. (2001). 'The predictive ability of Bromilow's time–cost model'. *Construction Management and Economics* **19**(8): 759–764.

Kumaraswamy, M. M. and Chan, D. W. M. (1995). 'Determinants of construction duration'. *Construction Management and Economics* **13**(3): 209–217.

Lyons, L. C. (1998) 'Meta - analysis: methods of accumulating results across research domains'. On-line publication —*The Eric Database.*

http://ericae.net/edo/ED339748.HTM, or
http://www1.monumental.com/solomon/MetaAnalysis.html

Mak, M. Y., Ng, S. T., Chen, S.E. and Varnam, M. (2000). 'The relationship between economic indicators and Bromilow's Time–Cost model: a pilot study'. *Proceedings of ARCOM 2000*, Glasgow.

NEDO (1988). *Faster Building For Commerce*. London, HMSO.

Ng, S. T., Mak, M. M. Y., Skitmore, R. M., Lam, K. L., and Varnam, M. (2001). 'The predictive ability of Bromilow's Time–Cost model'. *Construction Management and Economics* **19**(2): 165–173.

Sidwell, A. C. (1981). A critical study of project team organisational forms within the building process. PhD Thesis, Department of Construction and Environmental Health, University of Aston in Birmingham.

Walker, D. H. T. (1995). 'An investigation into construction time performance'. *Construction Management and Economics* **13**(3): 263–274.

Yeong, C. M. (1994). Time and cost performance of building contracts in Australia and Malaysia. Unpublished Master of Science Thesis, School of Building and Planning. Adelaide, University of South Australia: 288.

6

Net cash flows

INTRODUCTION

Construction project gross cash flows have attracted a great deal of effort from the research community, with attempts to illuminate their behaviour or to model that behaviour for predictive or simulation purposes. This work has resulted in a wide range of approaches and different mathematical techniques.

The same cannot be said for net cash flows. Here the research has been much more restrained and much less exploratory. The work has predominantly been reductionist, tending to shy away from empirical analysis. Models have been developed largely from assumptions and assembled rules, rather than from direct observation of project data.

This lack of attention is surprising, given that the need for understanding net cash flows is much greater, and the significance to the firm more direct. The relationship between gross cash flow modelling and the financial viability of the firm is often highlighted to justify cash flow research. This claim is hard to support on its own, but as the basis for understanding the net cash position of the firm then the claim becomes very strong. It is not the timing of costs and incomes on projects which effects the financial viability of a company, rather it is the relative timing of those payments.

The management of cash flow and the generation of positive rather than negative funds across a portfolio of projects is critical to the financial viability of any construction company. It is important to see the individual project in the context of that portfolio (see Chapter 7). However, before this can be done, it is necessary to develop models for individual project net cash flow.

Analysis can be undertaken from the top down such as the balance sheet approach of Punwani (1977), or from the bottom up with empirical analysis of project cash flow data (the more common approach).

Empirical analysis of net cash flows is difficult to achieve, because financial data is highly sensitive, perhaps more so in construction than in other industries, and is usually protected behind a wall of secrecy and policy. Very few companies are willing to reveal their innermost financial secrets, many will simply conceal the truth rather than make public the true nature of their financial dealings. While there is no reason at all for a firm to reveal the way it finances its business, this lack of willingness to participate in research in this area, goes a long way to explain the lack of empirical research undertaken, and therefore the relative dearth of alternative models for net cash flows.

Interest in net cash flows arose in the mid 1980s. As discussed in Chapter 2, interest rates at this time reached extremely high levels, with Australian ten-year government bonds at about 15%, and much higher rates for short-term funds. Such rates encouraged firms to explore ways to exploit the profit-making potential of the timing difference between moneys in and moneys out. Research was funded by industry, and projects were undertaken, to explore the behaviour of project net cash flow and its contribution to the organisation. This attention has since waned, due to the fall of interest rates to much lower levels, of about one third the peak rates for long-term funds.

This loss of focus is unfortunate. The true value of net cash flow research is not in the interest earning capacity inherent in the cash flow stream (although this is of course very important) but rather in its role in either financing the business or placing a drain on its resources. It is as a financial engine that cash flow management is critical, and it is in this sense that research is both needed and sadly lacking.

Henry *et al.* (2000) has established that suppliers play a significant role in financing construction and that providing this credit places them at risk while at the same time restricting their capacity to undertake new business. This is a crucial component of the construction industry, and a reason for improved modelling of net cash flow.

In the next three chapters, it will be demonstrated how net cash flow research can inform construction business management. This chapter provides the critical first step on that path, exploring the analysis and modelling of construction project net cash flows. It will be shown that net cash flows are highly individual but that the bulk of research has been built around assumptions of standard phase or weighted delays. This assumption has led to the failure of the research community to come to grips with the individual variation in net cash flow between projects. This individual variation highlights the potential of each project to contribute to the organisation and is an issue of fundamental importance in the industry, one which truly cuts to the heart of the financial viability of a construction company.

THE NATURE OF CONSTRUCTION PROJECT NET CASH FLOWS

Terminology

Cash flow (generally)

The general convention that the term 'cash flow', on its own, indicates the flow inward from the client to the contractor was introduced in Chapter 3. Correspondingly, the term 'outward cash flow' was selected for the payments out to contractors, suppliers and direct costs. The convention is specific to the building industry. Accountants would generally hold that cash flow concerns flows of cash in both directions. To avoid confusion the term used within this book for the combined cash flow is 'net cash flow'.

Component cash flows

Component cash flows are, within the context of construction project net cash flow, the individual inward and outward flows of cash on a project (Figure 6.1).

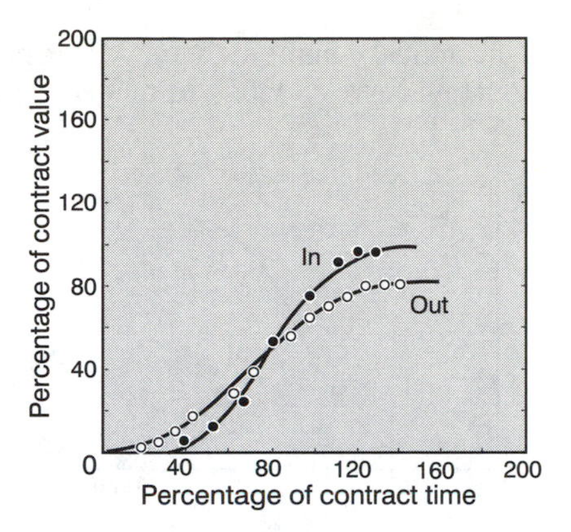

Figure 6.1 Typical component cash flows (Kenley and Wilson, 1989)

Net cash flow

The net cash flow for a project is the difference between the money flowing in to the contractor and the money flowing out (Figure 6.2). The important issue is the relative timing of the two flows. The term also has a common use in the property sector, for the calculation of the net present value (NPV) of the incomes and outgoings of an investment stream. This use does not have the same emphasis on the relative timing. The timing is not only important in the consideration of net cash flow, but is critical to understanding the financial performance of construction companies.

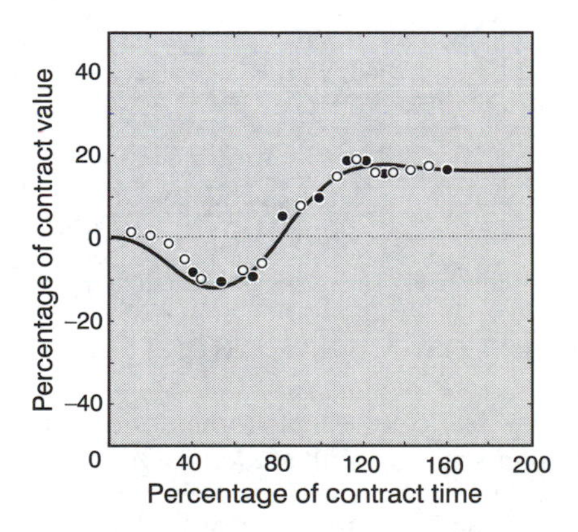

Figure 6.2 Typical net cash flows (Kenley and Wilson, 1989)

Diagrammatic view

Perhaps the easiest way to view the theoretical construct of this structured relationship is via a diagram. Figure 6.3 displays the path from commitment through outflow and inflow to net cash flow. The commitment curve is a theoretical concept and some believe that cost commitment should be modelled, others the income commitment, and maybe it should be a theoretical commitment of work in progress. Regardless, it may be viewed as an originating curve and may be modelled accordingly.

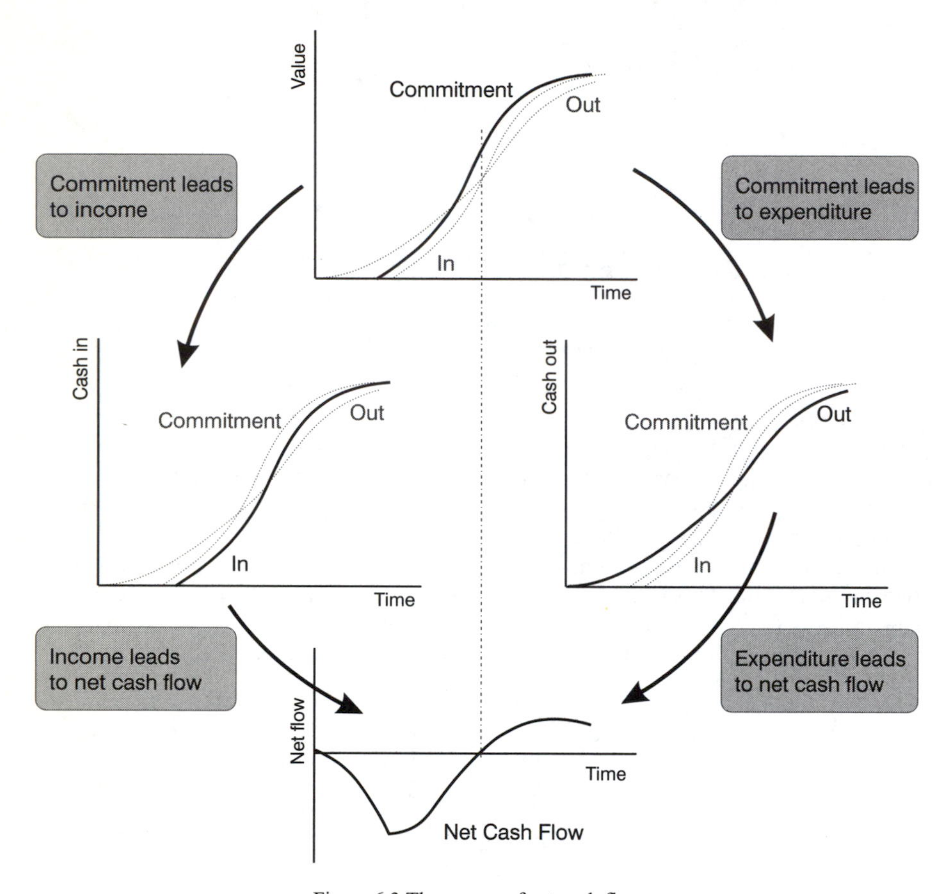

Figure 6.3 The source of net cash flow

ROLES IN NET CASH FLOW MANAGEMENT

The construction industry has a series of conundrums which make good dinner conversations or for light discussion with students. Questions such as:

- Is over claiming a variation fair game or fraud?
- Who owns the float on an activity on a schedule, the client or the contractor?

These questions can stimulate interesting debate, yet they have their serious side, sometimes emerging in disputes and other legal circumstances.

A question with a similar capacity to stimulate learning as well as being a very serious issue for the industry, is:

- Who owns the cash flow on a project?

To answer this question one must consider the project stakeholders.

The impact on the client

The client has commissioned the project, is paying the bills and generally has the right to only pay for work progressively, after it is completed. As the ultimate payer, the client (or their financier) has considerable power in the procurement agreement. The form of the relationship is obviously a factor, but most clients believe that they have significant ownership of the cash flow stream which arises through construction projects.

This can be seen quite clearly in the current emphasis on Construction Contracts or Security of Payments legislation and the related move to include in Head Contracts terms which require proof of payment to subcontractors and suppliers. In these circumstances, the client is taking a certain degree of ownership over the cash flow.

An argument can be made that the client should own the cash flow stream. They originate the cash flow stream, and are fully involved in it.

The impact on the contractor

In most contractual arrangements, the contractor has the most power to control the cash flow stream. They determine when they ask for money and they determine when they pay their bills. Whereas, no one questions the banks about the difference in timing between the payment of a commitment to the bank and their corresponding payment of their related commitments, there is a strong ethical concern (or cringe) with the role of contractors in managing their net cash flow. Manipulation of the cash flow stream, so common in the finance industry, is often seen as unwarranted and unscrupulous use of a position of power within the construction industry. Why should attitudes be so different in the construction industry? Perhaps the difference is that the banking system generally has safeguards in place to protect those with less power in the financial relationship.

A further argument can be made that contractors should own the cash flow stream. All payments in the cash flow stream go out to subcontractors and suppliers, through them, from the client. Their role is to manage the complexity of the project for the client; this includes the complexity of the payment system.

The impact on the supply chain

Existing payments systems impose heavily upon members of the supply chain beyond the contractor. When a builder fails, they pay a very heavy price for their involvement in the project, usually out of proportion to the amount of work they could do in one month. Ultimately these people do the work, and are paid by the client through the contractor. Henry *et al.* (2000) recognised the extent of this commitment, and studied the interaction between the provision of trade credit and the profitability of the suppliers. They concluded that profitability improved for those suppliers restricting access to credit, requiring guarantees of payment and maintaining tight credit control.

Another strong argument can be made that the supply chain should own the cash flow stream. All payments in the cash flow stream ultimately go to them, and in fact it can be shown (Gyles, 1992) that 80% of the cash belongs to this group. They do most of the work, and they should not have to wait to be paid. In many other industries they would be paid within a fixed term from when they delivered a product.

There is no correct answer

The question, who owns the cash flow, is rhetorical. It serves to illuminate issues rather than demand an answer. The reality is that all positions have some degree of truth and all players have a stake. Ownership of the cash flow stream becomes a power play. In the normal course of events, it is controlled by the contractors as they are generally in the position of greatest power. Government can act to change this through regulation, and major clients can place demands should they see fit, however only a limited range of specialist contractors and suppliers (usually manufacturers) are in a position to exert sufficient power in their supply chain relationship to control their cash flow.

Perhaps the most important issue is that the whole sector (including government) needs to rethink the cash management structures to ensure both adequate access to, and use of, the cash flow stream as well as protection of it.

PRACTICAL CASH FLOW MANAGEMENT

The net cash flow profile for a construction project looks something like that given in Figure 6.4, which is a diagrammatic representation of a smooth flow of cash over the life of the project.

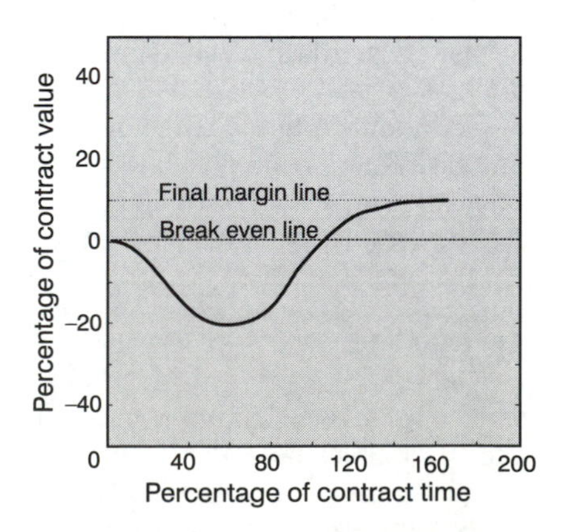

Figure 6.4 Idealised smooth net cash flow profile

Periodic cash flow

In Chapter 3 the periodic and cumulative nature of the flow of cash in from the client was discussed. The general principle is that there are lump-sum payments, usually monthly but empirical analysis suggests that this is not necessarily the case, and more frequent payments may be common. These inward cash flow streams are 'stepped'. As lump sums, these payments present problems for analog models, which prefer continuous change. There are various ways to represent continuous points for periodic cash flow. A typical stepped inward cash flow with joined line representations are shown in Figure 6.5.

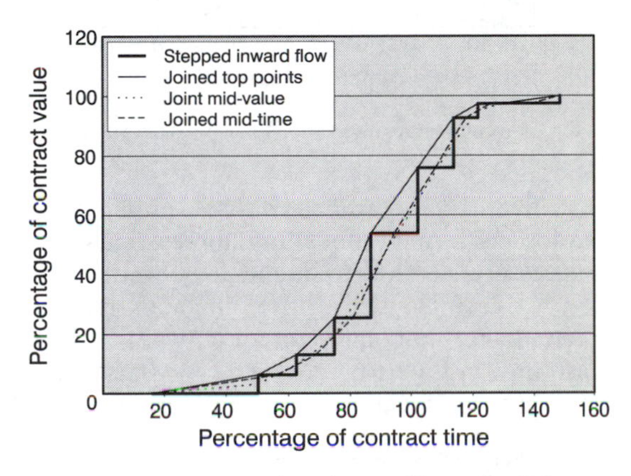

Figure 6.5 Representation of stepped inward cash flow

Continuous cash flow

The flow of cash out to subcontractors, suppliers and direct expenses such as labour, is much harder to specify. Some components of the cash flow will be discrete (lump sum), such as payments to subcontractors. Other components will approximate a continuous pattern, such as direct labour (weekly), many materials (for example, payment for concrete is often seven days from delivery) and sundry site expenses. Convention has it that the outward flow is represented as a continuum, although it is clear from the empirical evidence that much of the flow is tied into subcontract agreements which replicate the head contract and are therefore subject to the same lump-sum patterns.

Figure 6.6 shows the same project as Figure 6.5, but with the outward project cash flow instead of the inward flow. Note that the weight of the lines has changed to reflect the convention that the top points are used to represent this flow.

The hybrid net cash flow

Net cash flow is the balance between the inward flow and the outward flow. As a result it is ordinarily represented as a combination of the stepped inward cash flow and the continuous outward cash flow. For the project above, this would appear as shown in Figure 6.7.

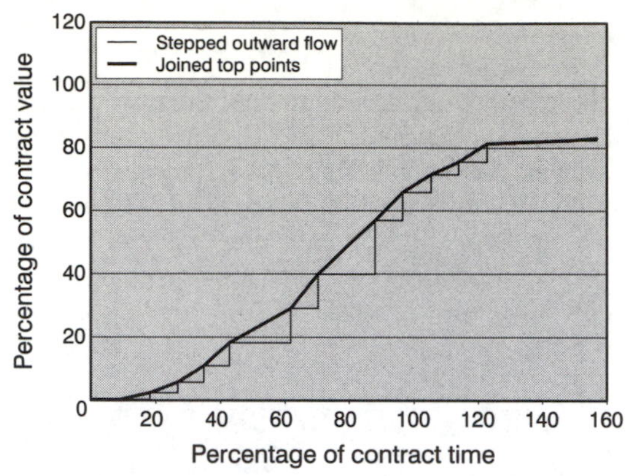

Figure 6.6 Conventional representation of stepped outward cash flow

This 'sawtooth' effect is very difficult to model. It is quite suitable for representing real data, but for any form of modelling, forecasting or simulation it cannot be represented. It is necessary therefore to smooth the inward cash flow to remove the jagged nature of the net cash flow. The three methods illustrated in Figure 6.5 combined with the continuous top-point trend for outward flow from Figure 6.6, are shown for the net cash flow in Figure 6.7. Nazem (1968) used the mid-points of the sawtooth when searching for an ideal reference curve for net cash flow.

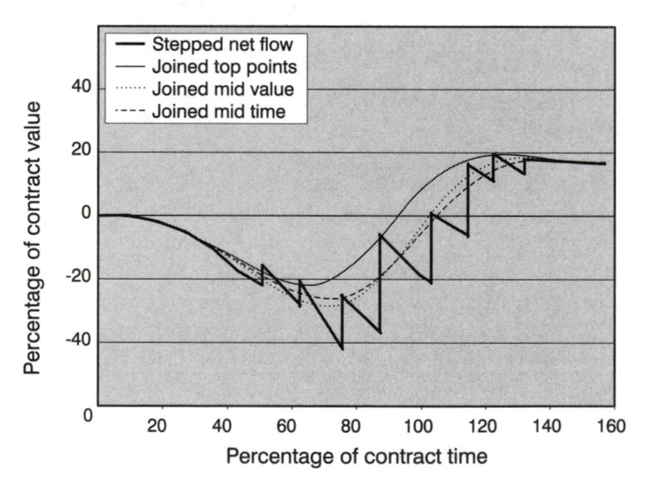

Figure 6.7 Representation of stepped net cash flow

THE THEORY OF MODELLING NET CASH FLOWS

Overview

Standard or ideal curves

In an early, seminal work, Nazem (1968) proposed a net cash flow model based on historic data, with the aim of discovering standard balance curves. He attempted to develop an 'ideal net cash flow reference curve' for use in predicting future capital requirements. Contractors do not undertake only one project at a time, therefore Nazem emphasised the overall requirements of the company, not the individual position for a project. He argued that an overlay of all projects would yield the capital requirements for a company over time. Nazem's proposal required that the ideal reference curve be derived as an average of a reasonable sample of projects. This method has not been successfully followed up, possibly due to problems in deriving such an average. Some construction companies very likely employ a similar technique as part of their management systems, to manage their net cash flow and encourage net cash outcomes which are to the long-term benefit of the company.

Balance sheet analysis

An alternative to forecasts using ideal reference curves was proposed by Cleaver (1971) who suggested that company operating (working) capital should be assessed at the balance sheet level. By this method ratios could be found which, when applied to turnover levels, would yield the amount of operating capital needed.

Such an approach was taken by Punwani (1997), who clearly identified from the balance sheet that firms generate working capital from turnover. Punwani analysed major construction companies from the UK and found that in most cases as turnover rose, their contribution from working capital increased. It was also found that firms reinvested these funds in other aspects of their business. Punwani's approach is discussed further in Chapter 7.

Weighted mean delays

The method of 'weighted mean delays' has become the dominant method for modelling net cash flows. In the absence of an ideal net cash flow curve, most research has concentrated on the method of weighted mean delays in order to develop a method for modelling individual construction project net cash flows. This method involves applying systematic delays to inward cash flow profile, in order to reckon the outward cash flow profile. The balance between the two is the net cash flow.

Peterman (1973) proposed an early model utilising standard delays, and this was followed by Ashley and Teicholz (1977) and McCaffer (1979). McCaffer refined the approach using forecast income schedules based on network analysis. These models did not use standard sigmoid (S) curves as their base, although both Ashley and Teicholz (1977) and McCaffer (1979) suggest that standard curves may adequately replace the more complex and expensive derivation of project income schedules. Trimble (1972) was of the opinion that the net cash flow is not sensitive to the shape of the standard curve chosen. This is a hint toward the future approach of researchers, who, finding data difficult to access, would develop net cash flow profiles of remarkably consistent profile, despite the originating curves.

Hardy (1970) utilised a system with gross cash flow curves derived from the application of built-up rates to a network schedule (PERT analysis) for the inward cash flow. He used this, and an outgoings curve inferred through applied systematic delays, to derive the net cash flow forecast for a project. This was then updated by reference to payments from the client, with appropriate allowance for delay. The delay system was utilised because:

> It is generally not practical to determine when the financial commitments of a project are actually settled because a supplier may service several projects, and a gross monthly cost category may be composed of a large number of invoices (Hardy, 1970).

The delay system resulted in '...no yardstick being available for comparison between forecast and actual payments' (Hardy, 1970).

Hardy's model was difficult to test because of the shortage of available cash payment data, thus the delay system was initiated. Hardy was able to perform some testing on hypothetical projects, obtaining an indication of the model's ability to be consistent while accommodating change, but not of its accuracy. The model was therefore largely reductionist, based on the results of estimated (and assumed constant) delays which could not be tested using empirical data.

Hardy had many doubts about his model. He felt that the following points should be made before the model could be put to practical use:

> 1. It is unlikely that any method will provide more than a reasonable guide to [net] cash flows from a contract.
> 2. Results will only be acceptable if the original cost commitment/time curve shows the correct trends with time.
> 3. If value is not proportionate to expenditure then it will be necessary to produce both cost commitment and income curves (Hardy, 1970).

These are important points, as they challenge the assumptions central to weighted mean delay models, but there has been little work undertaken to further explore the relationship between the timing of inward and outward cash flows.

Mackay (1971) used standard curves for the originating curves rather than curves derived from forecast work schedules. Subsequently McCaffer (1979) produced a comprehensive computer program which forecast construction project net cash flows using standard curves. Use was made of standard curves because he considered the preparation of work schedules (which were only as accurate as the schedule) involved complex and expensive analyses at a time when resources would be least available.

The systematic delay method used by McCaffer (1979) relied on the hypothesis that the value curve can be modelled by the use of standard curves, and that inward and outward cash flow curves can be modelled by the application of delay factors to the value curve (as in Figure 6.3). The value curve represents the certified value of the work, so the delay of payment from the client (due to contractual or other causes) gives the inward cash flow curve. The outward cash flow or cost curve relates to an equivalent cost value curve, as it represents the payment for work done (as compared to certified inward payments). The outward value curve is calculated from the value curve through applied factors.

The outward cash flow curve is then found from the cost value curve, as the outlay will usually occur after a period of delay which varies according to the outlay and project conditions. McCaffer used a method of weighted mean delays to derive the component curves from the standard value set at the commencement of this procedure.

McCaffer's procedure is similar to the simpler method used by Ashley and Teicholz (1977) who defined the inward cash flow curve as the earnings minus held retention, with allowance for lag. Similarly, the outward cost curve was derived from the earnings curve using specified lags and percentages of earnings.

The results of the weighted mean delay method have not been directly compared with the actual historical data for a project forecast. This is unfortunate because one observation by Mackay (1971) was that the selection of an appropriate originating (standard) curve did not greatly affect the net cash flow yielded. This result suggests that the systematic delay model is dependent solely upon the selection of appropriate delays, and not upon the selection of the originating curve.

Modelling component curves

Both the ideal reference curve and weighted mean delay models have limitations, one being that they use methods which yield consistent results regardless of the selection of originating curves. This is surprising, given the wide variability for cash flow curves and the large amount of variability in empirical data between individual project net cash flows. A model capable of adjusting to a wide range of variable profiles would be preferable. Such a model is unlikely to use polynomial regression of net cash flow data, as 'the regression analysis has failed to produce a convincing explanation of cash flow differences' (O'Keefe, cited in Kerr, 1973). Further research must return to the work of Jepson (1969) who suggested that 'generating' or 'component' curves (the inflow and outflow profiles) be used to derive individual project net cash flows.

Peterman (1973) illustrated the large variation possible between the net cash flows for various projects, and the derivation of the residual or working capital profile from component income and cost ogives. Similarly Nazem (1968), McCaffer (1979) and Neo (1979) illustrated the interaction between the component ogives and the residual.

In order to model the residual curve it is necessary to smooth the inflow ogive which is in general stepped due to the periodic nature of client payments. Figure 6.7 shows the complexity of this problem. Research into client cash flow ogives has generally done this by considering only the top points of each step (for example, Bromilow, 1978). This assumption is fair while the model has as its aim to forecast the client's commitment to pay, and where the amount shown for the appropriate stage of the project is the amount due to the contractor. However, it is subject to challenge for the contractor's net cash position, as the top points are a false representation of the project's average cash position. Nazem (1968) and Peters (1984) used the mid points of the inflow steps, as the data for their net cash flow analysis. Thus, the net cash flow model requireinward cash flows the actual outward data points (line A on

Figure 6.8) and the average of the range expressed by the inward data points (line B on Figure 6.8). This method is intended to give an approximation of the contractor's average cash position for a project. This was the method adopted by Kenley and Wilson (1989).

This method is open to challenge. Are the outflows really as smooth as has been suggested? If not, then they also should be modelled using averages. In turn, this would mean that the additional complexity of averaging points would incorporate a phase delay into the model, but would achieve little more.

The method that should be adopted really depends on the purpose for which the model is to be used and the properties of the project data modelled.

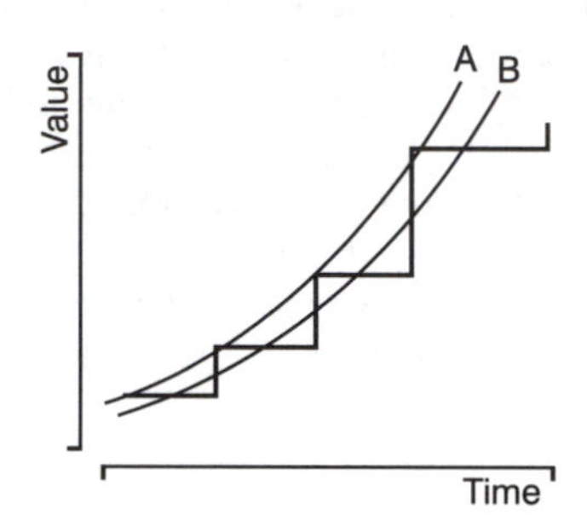

Figure 6.8 Part of a stepped net cash flow, showing (A) top point trend line and an alternative (B) mid-point trend line (Kenley and Wilson, 1989)

The Logit net cash flow model

Background

A model, based on empirical project data, has been developed that uses the actual component inward and outward cash flow data in order to form the residual model. This model adopted the Logit gross cash flow model (Kenley and Wilson, 1986) to model the component curves. When combined, the component curves form a net cash flow model reflecting the true net cash position for each project.

The gross cash flow model is described in Equations 6.1 to 6.3 below and is a family of curves which pass from (0,0) to (100,100) as percentages of project time and value. The model is capable of fitting both contractor's inward and outward cash flow.

The gross cash flow model, adapted from Chapter 3, can be re-expressed as:

$$\ln \frac{v}{V-v} = \alpha + \beta\left(\ln \frac{t}{T-t}\right) \tag{6.1}$$

which can be shown to be equivalent to:

$$v = \frac{V.e^{\alpha}\left(\frac{t}{T-t}\right)^{\beta}}{1+e^{\alpha}\left(\frac{t}{T-t}\right)^{\beta}} \tag{6.2}$$

$$\text{or } v = \frac{V.F}{1+F} \text{ where } F = e^{\alpha}\left(\frac{t}{T-t}\right)^{\beta} \tag{6.3}$$

and v is the percentage of value complete, t is the percentage of time complete, T and V are the total time and value, and α and β are constants. For convenience, T and V are normally expressed as 1.0 (ratio) or 100% (percentage).

For a gross cash flow model to be used for reckoning the net cash flow it must be possible to subtract an outflow from an inflow. This is not possible given the practice, accepted for the Logit gross cash flow model, of reducing all scales to the range of 0% to T=100%. Clearly 100% on the cash inflow curve is **not equal** to 100% on the cash outflow curve. For example total income may be $800,000 while total outgoings are $600,000 and yet both are defined as 100% on the ordinate. The ordinate values, for a given abscissa value, cannot be subtracted. Therefore a gross cash flow model must be adjusted to allow the percentages to be adjusted to real equivalents. This concept, which applies equally to the abscissa and the ordinate, is illustrated in Figure 6.9 which shows two component curves in Lorenz format (each curve is normalised using percentages and thus start and terminate at the same point).

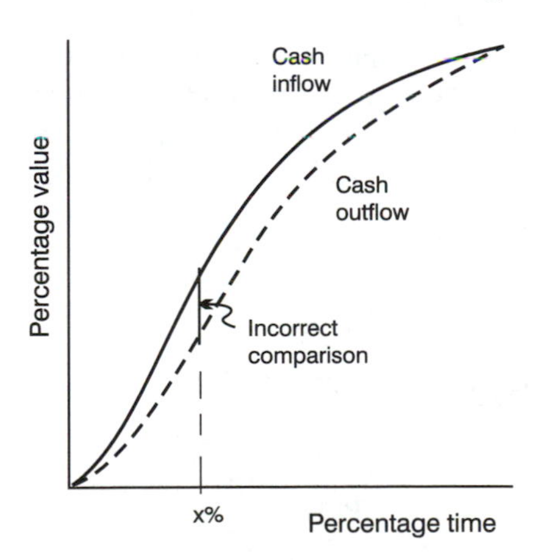

Figure 6.9 Component cash flow curves shown in Lorenz format

The percentage difference at time (t=x) is not equal to the real dollar difference, which is shown in Figure 6.10. The Logit construction project gross cash flow model is capable of being adjusted to use real dollar equivalents on shared axes.

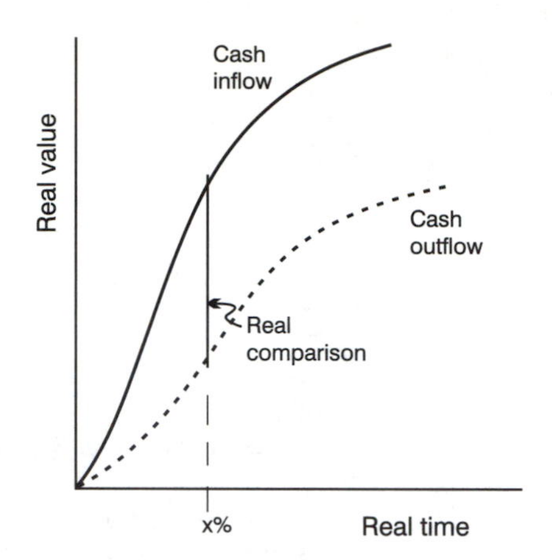

Figure 6.10 Component cash flow curves, expressed in real comparative terms

Combining component curves

It is possible to combine the separate equations for inward cash flow and outward cash flow components into a single equation. First, however, it is necessary to make an adjustment to the scaling and mathematics of the Logit cash flow model. A common base for the abscissa and for the ordinate, referable to both component curves, removes this problem and the value chosen for the base does not affect the functioning of the model. The suggested bases are 'initial contract value' for the ordinate (vertical axis) and 'initial contract period' for the abscissa (horizontal axis), as these are both known prior to commencement of the works.

The cash flow model therefore has a new base for the abscissa at the initial contract period, and becomes:

$$\ln \frac{v}{V-v} = \alpha + \beta\left(\ln \frac{t}{T-t}\right) \qquad \text{(6.1 repeated)}$$

where the maximum value of t is T:

$$T = \frac{\textit{total time passed at final payment}}{\textit{initial contract period}} \times \frac{100}{1} \qquad (6.4)$$

and the maximum value v is V where:

$$V = \frac{\textit{total cash flow}}{\textit{initial contract value}} \times \frac{100}{1} \qquad (6.5)$$

These values can also be placed into Equations 6.2 and 6.3 where the subscript s (e.g. V_s) is used to denote either inflow or outflow:

$$v = \frac{V_s \cdot e^{\alpha_s} \left(\dfrac{t}{T_s - t} \right)^{\beta_s}}{1 + e^{\alpha_s} \left(\dfrac{t}{T_s - t} \right)^{\beta_s}} \tag{6.6}$$

$$\text{or } v = \frac{V_s \cdot F}{1+F} \text{ where } F = e^{\alpha_s} \left(\frac{t}{T_s - t} \right)^{\beta_s} \tag{6.7}$$

If the residual at any time is the inflow less outflow, and if the subscript 1 (e.g. V_1) is used to denote inflow while the subscript 2 (e.g. V_2) is used to denote outflow, then the equation for the residual is:

$$r = v_1 - v_2 \tag{6.8}$$

Therefore using Equation (6.7):

$$r = \frac{V_1 \cdot F_1}{1+F_1} - \frac{V_2 \cdot F_2}{1+F_2} \text{ where } F_S = e^{\alpha_S} \times \left(\frac{t}{T_S - t} \right)^{\beta_S} \tag{6.9}$$

The mathematical expression for the component cash flow curves is not unique to the range $(0,0)$ to (T,V). Beyond this range, v decreases, whereas in reality value should remain constant. Therefore, if the component cash flow curves are to be superimposed, the basic net cash flow model must be adjusted to allow for the case where one curve reaches its maximum prior to the other.

An equation is not sufficient to give the value of r within the life of a project, so an algorithm is required. This allows v to have parameters which are conditional on the values of t relative to T. The algorithm may be written as follows:

if	$t \geq T_1$	then	$v_1 = V_1$
otherwise	$v_1 = \dfrac{V_1 F_1}{1+F_1}$	where	$F_1 = e^{\alpha_1} \times \left(\dfrac{t}{T_1 - t} \right)^{\beta_1}$
and if	$t \geq T_2$	then	$v_2 = V_2$
otherwise	$v_2 = \dfrac{V_2 F_2}{1+F_2}$	where	$F_2 = e^{\alpha_2} \times \left(\dfrac{t}{T_2 - t} \right)^{\beta_2}$
finally	$r = v_1 - v_2$		(6.10)

This algorithm (6.10) describes a family of curves which have the property of commencing at $(0,0)$ and moving by variable paths to r_f, a final value of r, which is the final contribution to profit and overhead. The algorithm describes the overall trend in working capital from the commencement of the project to the end.

Refining the model

The gross cash flow model was designed so that the curve would pass through the (0,0) and (100,100) points. However, in adapting the model for the purposes of the net cash flow model, it was found above that these constraints were no longer required. The final point of the curve, though still the final dollar value at final time, is expressed as a percentage of the initial contract time and value. A similar change applies to the way the model handles the commencement of the works.

The projects used in the gross cash flow analysis all had income streams which commenced relatively soon after their contract commencement date. Perhaps this was due to the management practices or the method of recording data of the companies involved. Furthermore, the model itself is not sensitive to early data, so it proved adequate to assume commencement at contract commencement for the gross cash flow analysis.

A net cash flow model must be more sensitive to the early stages of a project and the assumption that all gross cash flow curves should start at (0,0) is inadequate in a net cash flow analysis and the net cash flow model so far described is limited in its ability to handle the early stages of a project. Figure 6.11 illustrates a residual curve based on Algorithm 6.10. This curve shows that the model cannot respond to what may be considered a critical dip, or negative phase, at the start of the project. Many projects used for the net cash flow analysis were found to possess such a negative phase, and this observation supported the need to further modify the net cash flow model. On examination it was found that the data sets for the residual analysis occasionally showed extensive delays prior to first payments in either or both cash flow components.

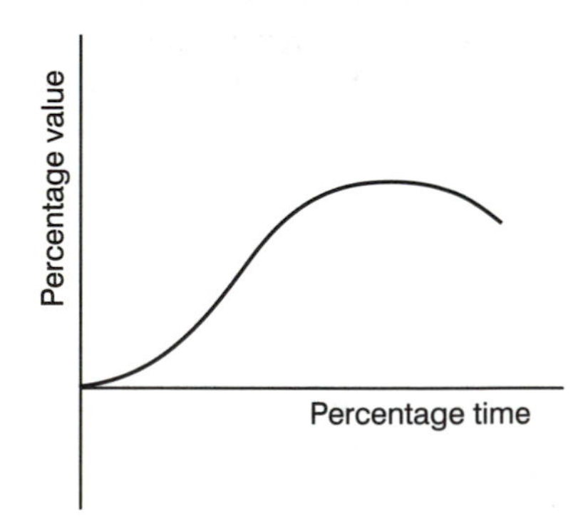

Figure 6.11 Net cash flow curve from algorithm 6.8

Therefore it was necessary to adjust Algorithm 6.8 to allow for the initial delays which might affect either or both of the component cash flow streams. In simple terms, the S curve for each component must commence just prior to payments for either component stream. This adjustment requires an initial phase delay and means that the time value of the S curve commencement may be different for each component.

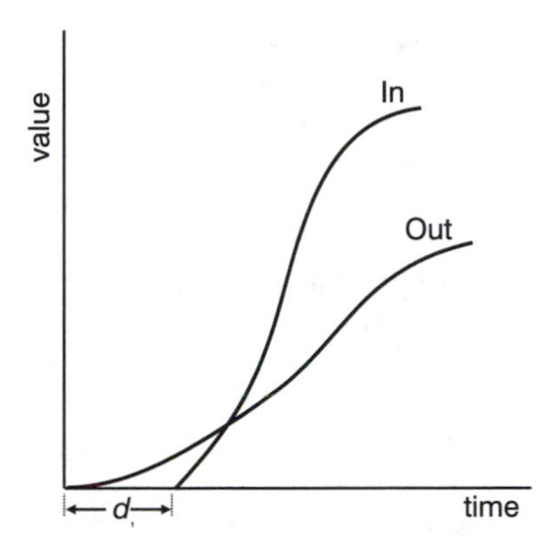

Figure 6.12 Component gross cash flow curves for a sample project showing initial phase delay

The concept of phase delays is illustrated in Figure 6.12, where d_1 is the delay associated with the income stream, and d_2 (the delay associated with the outgoings stream) equals zero.

In this way almost the complete ambit of possible component curves can be modelled, each may commence and end anywhere in relation to the other component curve. The balance curve which would result from the components shown in Figure 6.12 is shown in Figure 6.13. It has an initial dip due to the commencement of the outward cash flow trend line prior to that of the inward cash flow trend line.

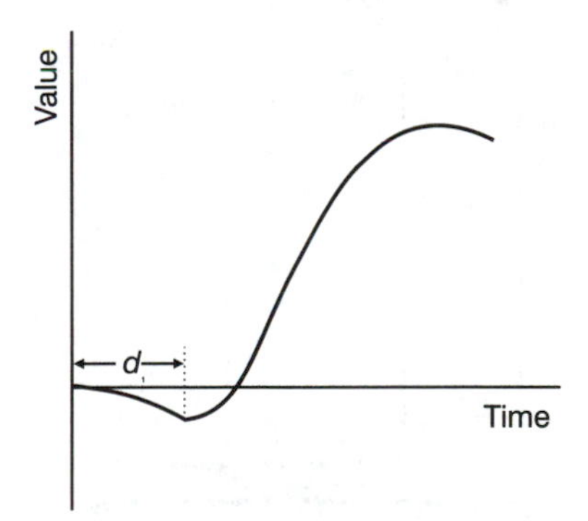

Figure 6.13 Net cash flow from Algorithm 6.10

Projects which have a delay prior to initial client payments can be accommodated within the model proposed above. Similarly a project which is delayed in its entirety can be modelled, as the origin is now effectively divorced from the commencement dates of the component flows (a factor which can be exploited in constructing organisational cash flow models). The Bromilow and Henderson (1977) industry standard gross cash flow model incorporated a fixed delay at the start of 10.7% (Balkau, 1975) and later Tucker and Rahilly (1982) used 1.9%.

The algorithm for the net cash flow model can be adjusted for delay through a twofold process. First the data is manipulated prior to the calculation of the α and β constants for each component gross cash flow so that the delay is effectively removed. Secondly, the constants so calculated are substituted into a revised model.

The α and β constants are calculated by adjusting the data by a factor to allow for the initial delay. Each data value is adjusted such that the abscissa has a new origin at the value of the shift required, while the end point remains unaltered. For the case of a 27% delay (of project time) prior to the effective start in the S curve, each data point abscissa is adjusted by the following formula:

$$X_{new} = (X_{old} - 27)\frac{100}{100-27} \tag{6.11}$$

In this situation, the original data range from 27% to 100%, and are therefore unsuitable for inclusion in the Logit model developed in Chapter 3, the revised data range from 0% to 100%, and therefore satisfy the requirements of the Logit gross cash flow model.

The general case adjustment is:

$$t' = (t-\theta)\frac{T}{T-\theta} \tag{6.12}$$

where θ is the imposed delay, t is the percentage of time passed, t' is the adjusted abscissa value, and T is total time as a percentage of initial contract time and is represented by Figure 6.14.

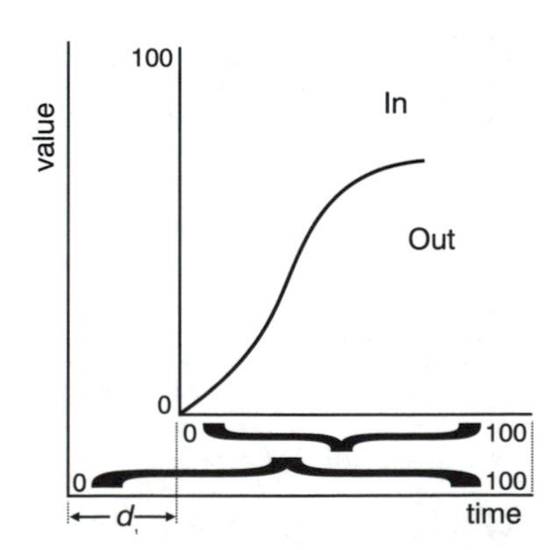

Figure 6.14 Adjusting the Logit model for the phase delay
(adapted from Kenley and Wilson, 1989)

Once the adjustments to the data have been made, the procedures for the derivation of the α and β parameters for the component gross cash flow curves may be followed.

Algorithm (6.13) for ordinate values for the revised gross cash flow model, incorporating a phase delay, is:

$$
\begin{array}{llll}
\text{if} & t \le \theta & \text{then} & v = 0 \\
\text{else if} & t \ge T & \text{then} & v = V \\
\text{otherwise} & v = \dfrac{V.F}{1+F} \text{ where } F = \mathrm{e}^{\alpha}\left(\dfrac{t}{T-t}\right)^{\beta} & & \\
\text{and} & t' = (t-\theta)\dfrac{T}{T-\theta} & &
\end{array} \qquad (6.13)
$$

Algorithm (6.14), for construction project net cash flows is:

$$
\begin{array}{llll}
\text{if} & t \ge T_1 & \text{then} & v_1 = V_1 \\
\text{else if} & t \le \theta_1 & \text{then} & v_1 = 0 \\
\text{otherwise} & v_1 = \dfrac{V_1 F_1}{1+F_1} & \text{where} & F_1 = \mathrm{e}^{\alpha_1} \times \left(\dfrac{t}{T_1 - t}\right)^{\beta_1} \\
\text{and if} & t \ge T_2 & \text{then} & v_2 = V_2 \\
\text{else if} & t \le \theta_2 & \text{then} & v_2 = 0 \\
\text{otherwise} & v_2 = \dfrac{V_2 F_2}{1+F_2} & \text{where} & F_2 = \mathrm{e}^{\alpha_2} \times \left(\dfrac{t}{T_2 - t}\right)^{\beta_2} \\
\text{finally} & r = v_1 - v_2 & &
\end{array} \qquad (6.14)
$$

where V is total value as a percentage of initial contract value, T is total time as a percentage of initial contract time, t' is adjusted abscissa value, t is actual abscissa value, θ is imposed initial delay, and α and β are constants derived from the data.

The calculation of v_1 or v_2 can be executed using the function in Table 6.1 using *Visual Basic for Applications*:

It has been found that this algorithm, which is ideally suited to computer applications, provides an excellent fit (considering the complexity and variability of the required profile) to the majority of project net cash flows. Six examples of project net cash flow data and the resultant fit from the model are illustrated in Figure 6.15 (page 181).

The code in AdjustedCCF() is generic and allows introduction of real data, percentage data or ratio data.

Table 6.1 Code for the function *AdjustedCCF*()

```
Function AdjustedCCF(TotalValue, TotalTime, StartValue, StartTime, DelayTime, Alpha,
Beta, ThisTime)
  Days = ThisTime - DelayTime
  If ThisTime <= DelayTime Then
    AdjustedCCF = 0
  ElseIf ThisTime < TotalTime Then
    TimeP = (Days * (TotalTime / (TotalTime - DelayTime)) / TotalTime)
    F = Exp(Alpha) * (TimeP / (1 - TimeP)) ^ Beta
    ValueP = F / (1 + F)
    AdjustedCCF = ValueP * TotalValue
  Else
    AdjustedCCF = TotalValue
  End If
End FunctionEnd Function
```

TESTING THE NET CASH FLOW MODEL

Data

The model was tested using two separate and different sets of data. The first sample S1 comprised data from 21 medium to large scale commercial and industrial projects, provided by a Melbourne-based construction group. The projects were constructed in the late 1970s to early 1980s, in Melbourne and some other major Australian centres.

The second sample S2 comprised data from five medium scale commercial and industrial projects, provided by another Melbourne based construction group. This group's turnover was about a quarter of that of the company supplying S1.

The data from S1 consisted of monthly totals for inward cash flows and outward cash flows. The figures therefore represented the monthly balance of an imaginary 'cash at bank' account for the project. This data is unusual as very few construction companies keep detailed financial records for cash transactions in the form required for this analysis.

The data from S2 consisted of monthly totals for expenditure commitment, and dates and amounts of cash receipts for progress claims. Commitment and actual cash flow cannot be modelled simultaneously, so it was decided to utilise and adapt the method of applied systematic delays developed by Peterman (1973), Ashley and Teicholz (1977) and McCaffer (1979). To this end, an adjustment was made to the timing of the payments outflow based on the assumption that subcontractors' and suppliers' accounts were settled an average of 45 days after receipt of account (an examination of a sample of project accounts indicated that this assumption represented a fair approximation). However, the labour figures (also supplied as monthly totals) were used directly because they were paid weekly. Each stream (subcontracts, materials and labour) was applied as separate amounts in the model to allow for the different timing resulting from the adjustments made. The resultant data may be found listed in percentage form in Kenley (1987).

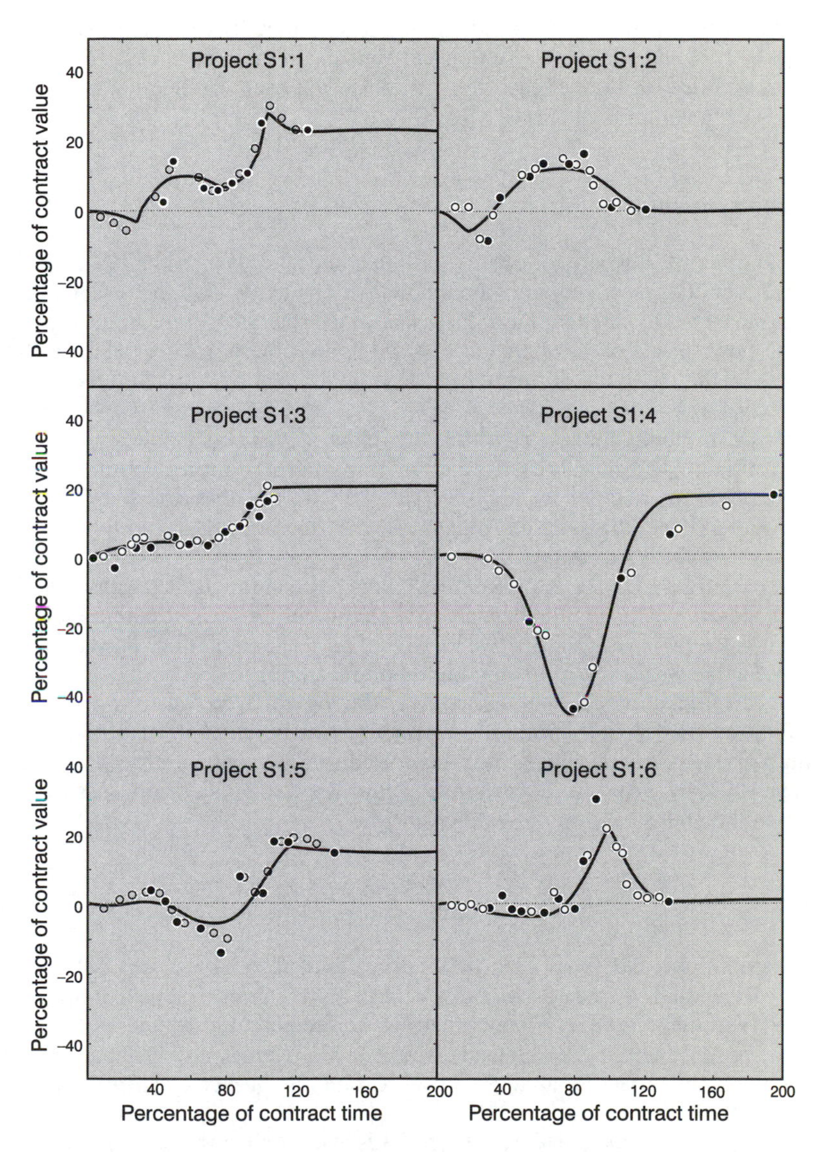

Figure 6.15 Examples of project net cash flow data and lines of best fit (Kenley and Wilson, 1989)

Exclusion of data at the extremes

A better fit for the gross cash flow model may be obtained by excluding data within the ranges 0–10% and 90–100% (Kenley and Wilson, 1986). This method also yields the best trend lines for the net cash flow model when used for calculating the parameters for each component curve.

Design and procedure

The conventions used when collecting and analysing the data in the net cash flow analysis are set out below so that comparisons may be made, and the tests repeated. Time was measured in calendar days. It was assumed that the Lorenz value ogive for gross cash flows based on working days would approximate that for calendar days; and thus so would the net cash flow curve. This approach ignores the effects of such discontinuities as holiday periods, strikes, delays, etc. This assumption was deemed necessary for simplicity and the results indicate that it was reasonable.

The initial (0%) time value was taken to be the date of contract commencement, or the date of commencing work upon the site, whichever came first. This is a more flexible approach than that taken for the gross cash flow model, due to the inclusion of the allowance for initial delay.

The base figure for 100% of time was taken to be at the initial contract period. That is, the interval between the date set in the contract for project commencement, and the date set down in the contract as the date of project completion. The base figure of 100% of value was taken to be the initial contract value.

The end points of the component cash flow ogives were taken to be at (T, V) where T is the final project duration expressed as a percentage of initial project duration. Similarly V is the final value expressed as a percentage of initial contract value. The 100% base for inward cash flow is equivalent (on both the abscissa and the ordinate) to the 100% base on outward cash flow.

Results

The outward component gross cash flow curves tended to have lower SDY values than the corresponding inward component cash flow curves. Table 6.3 shows the average SDY values for the component curves before and after removal of outliers; outliers are projects with SDY values in excess of 6% (Kenley and Wilson, 1986) on either component curve. Sample S3 is S1 excluding outliers, sample S4 is S2 excluding outliers.

Six projects were excluded as outliers. This is more than would be predicted by the results of the gross cash flow exclusion analysis. Interestingly all but one of the outliers was excluded by the inward cash flow component SDY percentage. This suggests that outward cash flows are a better fit to the model than inward cash flows, a conclusion supported by the lower mean sample SDY figures for outward cash flow in Table 6.2.

Table 6.2 α and β for S1 and S2 with associated SDY values

S1						
In				Out		
α	β	*SDY*		α	β	*SDY*
1.74	1.53	3.84	1	1.02	1.47	2.75
1.18	1.20	1.47	2	0.91	1.27	2.54
1.56	1.34	2.54	3	1.57	2.19	2.49
–0.96	1.18	*6.01	4	–0.29	1.42	*6.64
0.73	2.33	6.88	5	0.53	3.87	2.98
1.43	2.53	3.91	6	0.61	2.08	3.41
0.93	2.33	2.09	7	1.68	2.59	2.22
0.12	1.18	1.70	8	0.34	0.84	*10.87
0.47	2.01	2.85	9	0.40	1.85	2.02
–0.28	1.64	3.69	10	1.40	2.30	3.47
1.54	2.13	3.83	11	1.61	1.79	2.51
–0.15	1.66	1.37	12	–0.03	1.38	2.87
–0.45	1.09	3.34	13	–0.27	1.36	3.17
–0.43	0.87	3.37	14	–0.54	2.20	1.22
0.45	1.30	3.31	15	–0.02	1.47	*3.07
–0.17	1.65	4.28	16	–0.41	1.74	2.55
–0.62	1.38	*6.97	17	0.23	2.23	3.80
1.11	1.76	2.53	18	0.85	1.49	1.60
0.80	2.37	2.83	19	0.21	2.37	4.71
–0.96	1.34	2.96	20	–0.64	1.90	2.55
–1.31	1.43	4.77	21	–0.97	1.93	2.96
S2						
In				Out		
α	β	SDY		α	β	SDY
0.12	2.03	2.02	1	1.40	2.52	3.19
-0.52	4.42	4.64	2	2.71	3.99	2.34
-0.28	1.47	2.83	3	0.47	1.55	1.43
-0.78	1.50	*7.75	4	0.12	1.60	5.37
0.28	0.57	*12.90	5	2.07	2.51	2.21

* Possible outlier

Table 6.3 Mean sample SDY

Sample	Flow	Mean flow SDY	σ
S1	In	3.69	1.50
	Out	3.35	2.06
S2	In	6.03	4.43
	Out	2.91	1.51
S3	In	3.29	0.86
	Out	2.71	0.78
S4	In	3.16	1.34
	Out	2.32	0.88

Table 6.2 shows the component gross cash flow SDY values for each of the projects examined. It also includes the α and β values for each component curve. These α and β parameters were entered into the algorithm 6.14 to derive the net cash flows for each project. Figure 6.16 shows the two component curves for the sample project displayed at the start of this chapter, and the resultant net cash flow curve shown in Figure 6.17.

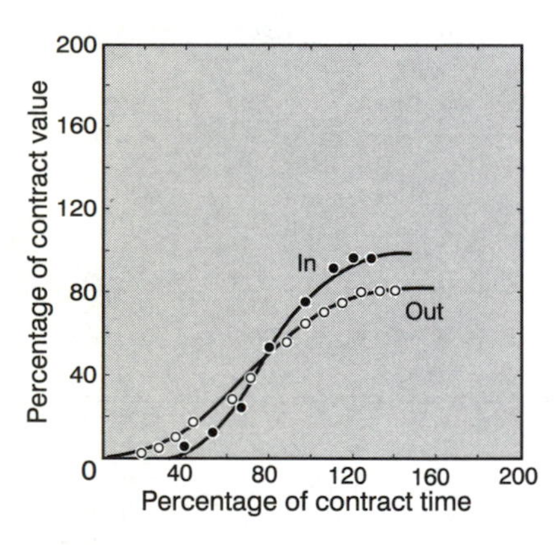

Figure 6.16 Component cash flow

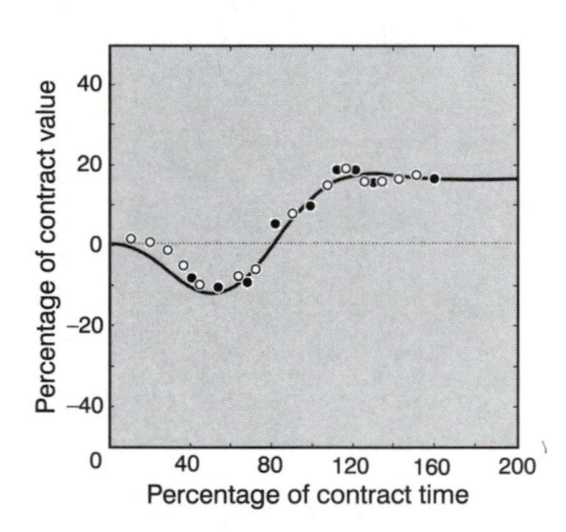

Figure 6.17 Net cash flow

A more detailed view (Figure 6.18) can also be taken of this fairly typical project showing the two component curves, the inward cash flow curve adjusted to mid-value-points, and the resultant net cash flow curves both by top-points and by the mid-points. The true net cash position can be expected to be between the two net cash flow curves displayed.

The net cash flow model was found to be an extremely good approximation to the trend line for the net cash position during the life of the project, for those projects with statistically acceptable SDY levels for the component curves. Thus the model was found to be acceptable for 75% of projects. Even for those projects excluded by the analysis the fit was sometimes acceptable. This is an excellent result for the residual model. The cause of failure of the model to fit the data can often be identified from the component curves. Frequently the cause was due to irregularities in the originating curve profiles, resulting in unmanageable net cash flow profiles.

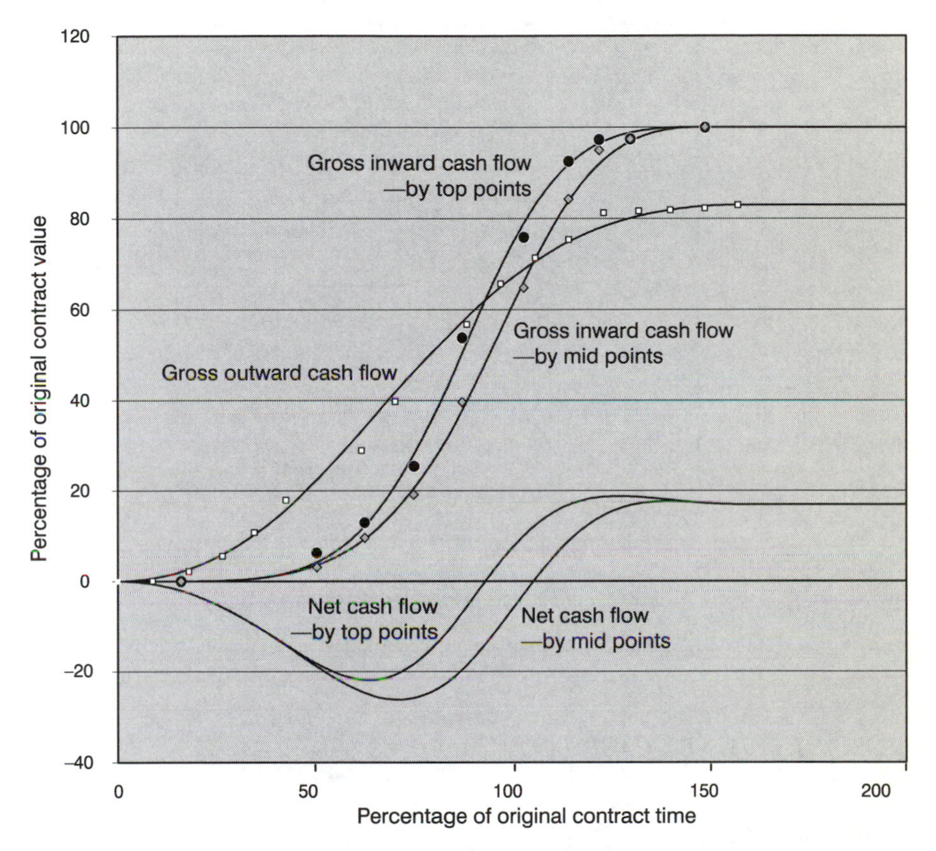

Figure 6.18 Detailed view of the sample project

The curves which result from the model are of widely varying profiles, and do not conform to a standard shape (such as an S shape) let alone a standard construction project net cash flow profile. Figure 6.19 illustrates the extent of variation in the profiles for projects accepted by the analysis. This figure presents a telling argument for an individual approach to cash flow modelling.

The model was tested by comparing the fit, by eye, of the net cash flow curves generated, to the originating data. Six of the projects analysed are shown in Figure 6.15 (page 181).

The Logit net cash flow model has proved to be successful for the post hoc examination of construction project net cash flows. The model is highly flexible and adaptive to the profiles of individual projects.

The results clearly demonstrate the wide degree of variation in profile between projects. Those with both component curves having a standard deviation of the estimate of Y (SDY) value of less than 6% (75% of projects examined) provide an excellent fit to the model. However, the model has only a limited ability to fit aberrant projects.

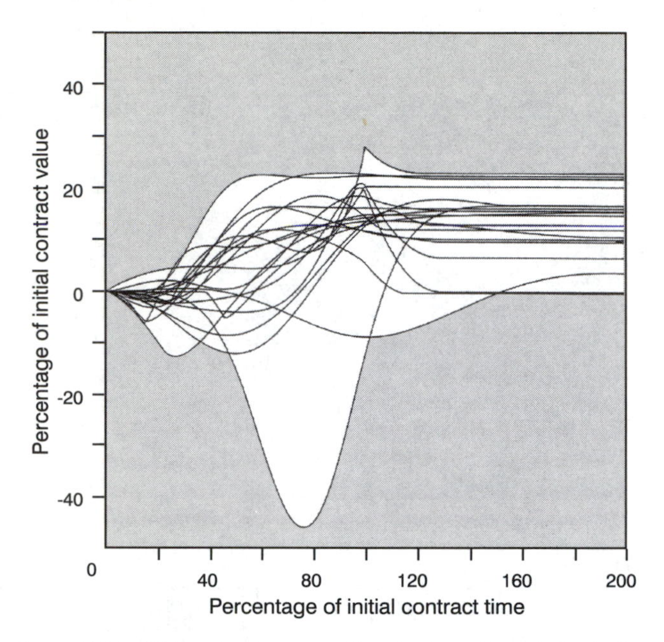

Figure 6.19 Multiple net cash flow profiles showing wide variation in form (Kenley and Wilson, 1989)

ALTERNATIVE NET CASH FLOW MODELS

Ashley and Teicholz

It is easy to overlook the importance of the work of Ashley and Teicholz, because their's is a small paper criticised by their peers, yet this would be a mistake as their methods and particularly their equations have been widely accepted in the research community. Ashley and Teicholz (1977) proposed the use of standard curves, but resorted to deriving their model from priced schedules, as mentioned in Chapter 3. They then developed a model for deriving the payments curve from the earnings curve, using the following equation:

$$P_t = \left(1 - \frac{R}{100}\right) E_{(t-b)} + M_t \text{ if retention is held, and} \quad (6.15)$$

$$P_t = E_{(t-b)} + M_t \text{ otherwise} \quad (6.16)$$

in which P_t is the total payment in month t, R is the retention percentage, $E_{(t-b)}$ is the earnings in month $(t-b)$ where b is the billing lag in months, and M_t is the mobilisation payment in month t ($M_t < 0$ for repayment). Their cost curve was made up of components:

$$C_t = \sum_{i=1}^{n} C_{t_i}, \text{ where} \tag{6.17}$$

$$C_{t_i} = K_1(E_{(t-b_1)}) + K_2(P_{(t-b_2)}) + K_1(C_{(t-b_3)}) + D_t \tag{6.18}$$

in which C_{t_i} is the cost amount of curve I in month t, K_1, K_2, and K_3 are fractional coefficients, b_1 b_2 and b_3 are time lags, in months, D_t is a lump sum or uniform amount spread to month t and C_t is the total of all costs in month t. N is the number of cost curves.

Kaka and Price

Kaka and Price (1991) have made a significant contribution toward investigating and modelling the properties of net cash flows on construction projects and in developing an organisational financial system for contractors. Kaka adopted the Kenley and Wilson (1986) Logit cash flow model as the underlying cash flow engine to drive this research which seems to follow closely the method outlined by Ashley and Teicholz.

In an earlier work, Kaka and Price (1991) tested the question raised by Kenley and Wilson (1989), regarding the extent of variability of individual net cash flow curves. They pursued an ideal net cash flow and concluded that this was not possible, but this was a conclusion tempered with reasonable estimates of net cash flow when comparing actual data with fitted data.

Their model used cost commitment curves to predict cash flows for individual projects. They compared the net cash flow of individual projects with a method for deriving cash flow which was based solely on the cost schedules rather than both the cost and value as proposed by Kenley and Wilson. This is in fact a method similar to the weighted mean delays discussed previously. Their contention was that the basic method of weighted delays required adjustment for the unbalancing effects of front-end loading of rates for profit motives or to correct measurement errors, as well as the effects of different subcontractor teams. To assess risk, the model also included random components on items such as inflation rates for cost escalation during contracts. The following components constructed the net cash flow model.

Simulated cost curve

This used the Logit model derived from historic data and adjusted to current dollar values, to derive an estimate of cost commitment. The cost commitment model was used as the basis of the entire model.

Table 6.4 Time lags for selected cost categories (Kaka and Price, 1991)

Cost type	Number of months delay				Percentage of total cost
	0	1	2	3	
Labour	100	0	0	0	15
Plant	0	20	60	20	5
Subcontractor	0	50	50	0	45
Materials	20	70	10	0	25
Site Overheads	100	0	0	0	10
Total cost delay	30	41	28	1	100

Cash out curve

The actual cost commitment curve was converted to cash out using the method of weighted mean delays which they called selected time delays. Table 6.4 shows Kaka and Price's time lags for their selected cost categories. These indicate the time period between the purchase of materials and payment and recognise that materials have a stock period (the time between purchased and use of that material). These figures provide a probability of inclusion in a given month after a given delay.

This table can be used to derive a table for converting the cost commitment curve to cash out. Kaka and Price provided the following sample table for this conversion (Table 6.5).

Table 6.5 Applying the time lags to derive cash out (Kaka and Price, 1991)

Cost curve (£)	Number of months delay				Total cash-out curve
	0	1	2	3	
0	0				0
50	15	0			15
100	30	20	0		50
160	48	41	14	0	103
230	69	65	28	1	163
1		94	45	1	–
1			64	2	–
				2	

Value curve

The derivation of the value curve (cash in) depends on the form of contract, and differs for fixed contracts, fixed contracts adjusted for inflation and cost–plus contracts.

- Fixed contracts: the cost curve is adjusted for the expected inflation rate and converted to a value curve.
- Fixed–adjusted for inflation: the cost curve is adjusted by the simulated actual rate.
- Cost–plus contracts: value curves are calculated from the simulated actual cost curve.

The detail of these conversions is not clear, however Kaka and Price provide the equation (6.19) for adjusting for mark-up. The reasonable assumption here is that there is a clear and direct relationship between the cost and the value—through the percentage mark-up applied by the contractor.

$$adjM = \frac{M.C}{C - Pr} \tag{6.19}$$

where M is the mark-up, C is the total cost of contract and Pr is the premium cost.

Front-end loading

The deliberate inclusion of front-end loading by Kaka and Price is very interesting (because the practice is ethically questionable) but not unusual.

> Unbalancing of a bid occurs when a contractor raises the price on certain items and reduces the prices of others so that the bid for the total job remains unaffected. The most effective type of unbalancing is front-end loading. The main reason for front-end loading is to shift the project financing from the contractor to the owner by increasing the unit prices on early items and decreasing the unit prices on later ones. This influences the value curve significantly (Kaka and Price, 1991).

This approach was originally proposed by Ashley and Teicholz (1977), the pioneers of the weighted delay approach, who proposed a linear unbalancing model. Their reasons for including unbalancing were to replicate contractor behaviour:

1. To shift the project financing from the contractor to the owner by increasing the unit prices on early items and decreasing the unit prices on later ones.
2. To properly distribute the fixed costs of the project when an error in the owner's quantities is detected.
3. To increase the total profit of the project by placing the higher unit prices on items expected to exceed the owner's estimate of work quantity.

Their linear unbalancing model required the multiplication of sums by the multiplicative factor applicable at a given time period (Figure 6.20). Kaka and Price favoured an alternative approach to solve the problem that the cumulative total varied depending on the size of the sums and that the factor had to be adjusted to ensure the total reached unity. They instead adjusted a difference from the totals for months from the corresponding months at the other end of the contract.

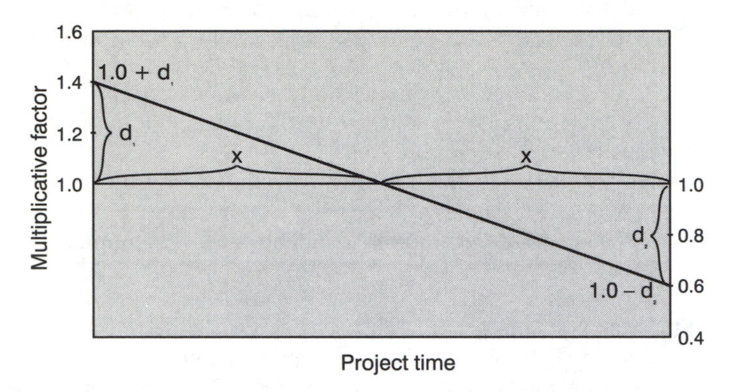

Figure 6.20 Linear unbalancing model (adapted from Ashley and Teicholz, 1977) (reproduced with the permission of ASCE Publications)

Materials on site

To allow for the pre-purchase of materials, to be paid for by the client, Kaka and Price provide a table of stock periods in their model. This table works in reverse, displaying the cumulative probabilities of stock periods in percentage of total material (Table 6.6).

Table 6.6 Example of stock periods (Kaka and Price, 1991)

	Stock period (months)		
	0	–1	–2
Percentage of total material	50	40	10

Retention

Ashley and Teicholz's method was once again adapted, from Equations 6.13 and 6.14:

$$\mathit{Effective\ retention\ rate} = \mathit{retention}\left(1 - \frac{SC}{V}\right) \qquad (6.20)$$

where *retention* is the contractual retention rate, S is the percentage of subcontract cost to total cost and V is the total value of the contract. Repayment is assumed to be paid 50% on completion and 50% at the end of the maintenance period.

Testing

With the net cash flow being the balance between the cash out and the cash in, Kaka and Price were able to compare real net cash flow data with their model on five projects. They tried hard to get more data, but encountered the problem of confidentiality which restricts this area of research. They concluded that their approach produced reliable forecasts for the limited sample used on a project by project basis. They did not reach the same conclusion from the results of drawing an average of the five projects as an ideal curve, which they found to be unsatisfactory for predictive purposes.

Kaka's improved model

An improved model was developed by Kaka (1996) to develop more flexible and accurate cash flow forecasting. Using a combination of greater detail in breakdowns (increased categorisation) and risk analysis (stochastic approaches) the revised approach represented a move away from the nomothetic ideal curve models toward more idiographic modelling of individual projects. The approach was to derive gross value curves for categories of expenditure and to derive the corresponding cash out curves individually. The stochastic approach comes through developing a range for two extreme cases for payment of retention moneys.

The model is heavily focussed on the reality of monthly payments, with adjustments for part-month commencement. It also explicitly recognises for the first time, the importance of different payment streams, being the measured work, the preliminaries and the materials on site. This led to a revised formula for mark-up:

$$P_m = \frac{(1+P)C - C_{pr}(1+P_{pr})}{M_c} - 1 \qquad (6.21)$$

where P_m is the overall mark-up applied to the cost of measured work, P is the overall mark-up applied to total cost, C is the total cost of the contract, C_{pr} is the total cost of preliminaries and M_c is the total cost of measured work.

It might be easiest to summarise the approach, as the method and terminology can become confusing.

- Derive **cost commitment curves** for separate components of the work (uses Logit model), with the components being own labour, materials, plant, labour only subcontractors, labour and materials subcontractors, nominated subcontractors/suppliers and site overheads.
- Derive monthly cash out for the components, including adjustment for discounts on payment, timing of payments (delays), payments for materials and retention.
- Derive the **cumulative cash out** for the project by summing the components.
- Returning to the cost commitment curves, derive the cumulative cost commitment. From this, adjust for materials on site, adjust for front-end loading and then apply mark-up according to the categories of measured work, the preliminaries and the materials on site. The sum of these components is the **cumulative cash in**.

The other major improvement in the model is the incorporation of risk through accepting ranges for data input into the model. Factors incorporated were cost variances, duration overrun, variations and under measurement.

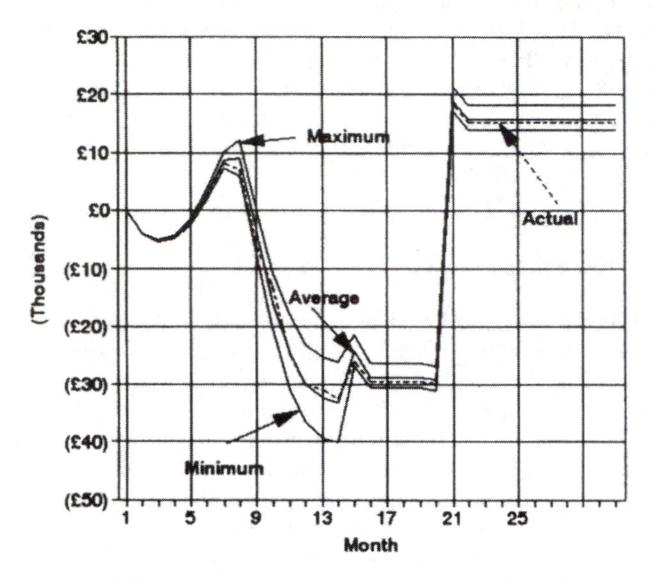

Figure 6.21 Average, maximum, minimum and actual cash flows obtained in second test (Kaka, 1996: 42, Figure 4)

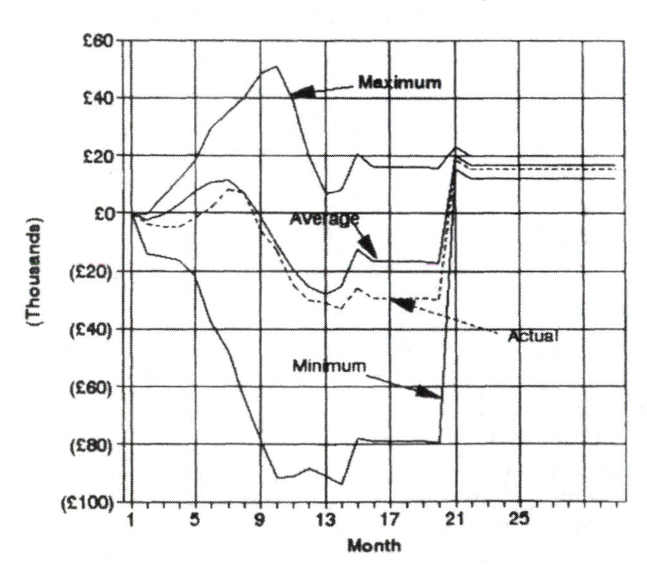

Figure 6.22 Average, maximum, minimum and actual cash flows obtained in third test (Kaka, 1996: 42, Figure 5)

Testing the model

Kaka tested the model and its ability to handle different profiles. Two charts illustrate the results, reproduced here in Figures 6.21 and 6.22. These show a wide envelope of net cash flow, but a very limited range of profile form. The extent of coverage is necessary and powerful, however, the limited scope of form reflects the highly structured construction of the model. This is not consistent with the results of Kenley and Wilson (1989) and it is possible that they pertain to a specific contractor and its circumstances, rather than being a general model. It may be noted that more variability in the profile had been obtained by Kaka (1995) previously, but this difference is not explained.

Navon

Navon (1995) did not really construct direct models for project net cash flow, but rather built computer-based systems based on a combination of integrative cost–schedule systems and the Ashley and Teicholz (1977) approach to net cash flow. However, Navon argued for the use of resources as the driver for the source cash flow. While not unusual in cost–schedule based systems (see page 130), such an approach is unusual in net cash flow and organisational cash flow models.

Navon's assumption is that the net cash flow can be driven by a cost profile (from which a marked-up income profile may be derived) obtained by linking resources and activities using an Activity–Resource–Linkage file. This is a project specific table drawn from a database of Activity–Resource relationships, called the general resource/activity allocation database (GRAAD). The system is, in fact, a relational model of Activity–Resource relationships. Navon identified that Activities link to Resources through cost items, which could be classified as follows:

1. A one-to-one relationship means that each activity involves only one cost item and that the cost item is unique to this activity; hence all the resources constituting the cost item can be allocated to the activity.
2. A one-to-many relationship means that two or more cost items relate to a single activity…all the cost items can be allocated to the one activity.
3. A many-to-one relationship is the opposite of the preceding one, i.e. two or more activities relating to a single cost item. Each of the cost item's resources must be divided among the activities.
4. The many-to-many relationship is a combination of points 2 and 3…The mapping problem (connecting cost items to activities) and the resource allocation to activities in this case are obviously more complex. (Navon 1995).

Navon's integrative model relies on the assumption that cash flow forecasts based on resources will achieve the highest accuracy. The typical resource allocation is best illustrated by means of an example and Navon provided the example of concrete pouring (Table 6.7).

The specifics of the database model are not relevant here, but the process of derivation of the income flow curve and the subsequent conversion to a net cash flow is relevant.

Table 6.7 An example of the General Resource/Activity Allocation Database (GRAAD) (Navon, 1995: 504, Table 1)

Activities		Resources	
Activity identifier	Description	Code	Description
		1225	Ready-mixed concrete
		1210	Cement
		1120	Gravel
		1110	Sand
22	Concrete Pouring	2100	Labour
		3122	Mixed usage
		3133	Conveyor usage
		3135	Crane usage
		3138	Pump usage

The system projects costs, including time lags and calendars. Navon uses a simple adjustment of two months or more for the lag between forecast cash flow and actual payments, being varied by the type of cost, direct, subcontracted, etc. Calendars are provided to handle the vagaries of holidays and the differing religious composition of crews (an issue of varying regional significance). There is more sophistication with the management of stock, although this has a distinctly civil engineering flavour and may not be particularly relevant in a commercial construction context.

Calculation of the income flow closely resembles Ashley and Teicholz's model, with adjustments for overheads and profit (mark-up) and then retention and billing period. Equation 6.13 is used also by Navon.

Navon's model is highly idiographic, being based on the actual project schedule, the actual bill of quantities and the relevant project contract conditions. However, there is no attempt to develop a model for the component curves. This is a clue to the direction of Navon's work which develops a company-level cash flow management model based on the cash flow system discussed here. Company level systems are discussed further in Chapter 7.

Khosrowshahi

The compound curve model developed by Khosrowshahi (1991) and discussed in Chapter 3 is used to form a model for net cash flow. Unfortunately, beyond the base of the combined Control module, Kurtosis module and Distortion module, the mechanisms have not yet been published. To Khosrowshahi (2000), cash flow is simply the difference between INs and OUTs with or without taking the interest charges and gains into account. The INs are forecasted by using the model and the OUTs are based on percentage of income incorporating the profit, or costs are phased out with certain delays. However, the two cash flows are distinguished by including the money lost or gained through interest gains and interest payments for positive and negative funds respectively (for each period separate rates are identified for both interest gains and charges).

This model is in the experimental stage at present, having been developed to explore financial risk management and cash flow management strategies, which will be discussed further in Chapter 7.

NET CASH FLOW PERFORMANCE MANAGEMENT

Benchmarking methods

There is an apparent interest in working capital management which was brought about by the high interest rates of the 1980s. The net cash flow model enables the establishment of standards of project net cash flow management, and for comparisons to be made between project performance and standards.

The management performance measured by the net cash flow model is the management of working capital gained through, or used by, operations on a project by project basis. It only remains therefore to set appropriate standards of construction project net cash management. The issue of performance management in working capital generation through operations at an organisational level is dealt with in Chapter 7.

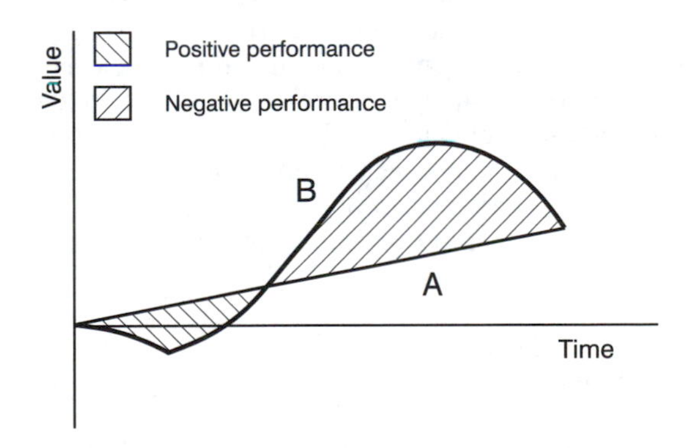

Figure 6.23 Project cash management performance, measured against a simple linear standard

If a project was being managed on a 'cost plus fee' basis, then the manager would pass on all costs to his client—plus a fee. His net cash flow would ideally be a straight line representing a fixed percentage of the value of work in progress. Thus one standard for the management might be a straight line from the origin to the final margin. This is line A on Figure 6.23, and an example project, line B, would by net cash flow profiles for most projects. This dip is a result of expenditure being incurred for which income cannot standard A appear to be poorly managed early in the project, and well managed for the latter two thirds.

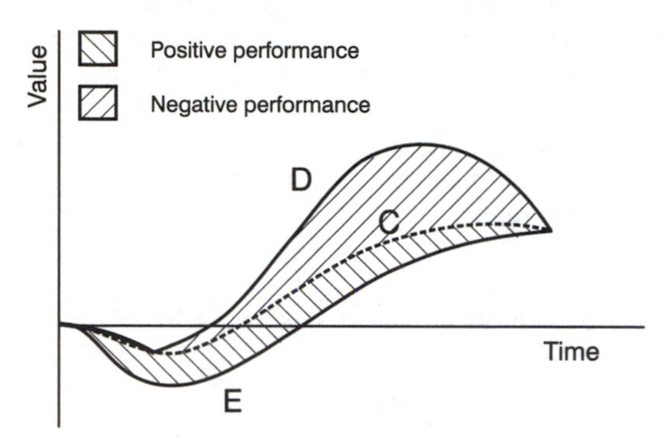

Figure 6.24 Project cash management performance, measured against a standard cash flow profile

In practice a straight line is unlikely for construction project net cash flow. A company may wish to set another profile for its standard of performance. It may be recognised that it is not possible to remove the initial dip arising immediately before first payments, as possibly the costs of setting up a project are not directly chargeable. Furthermore, it may be argued that good financial management will result in a greater proportion of money coming in than going out, and thus the later stages in a project will yield a net cash flow in excess of the line A in Figure 6.23. This being so, a standard may be set which incorporates these features, such as line C on Figure 6.24. Such standards could be set by management participation. The project D on Figure 6.24 would be seen as a successful project, while project E clearly has failed to meet the standard set. This is so even though the projects illustrated in these examples yield the same final contribution to overhead and profit.

This method provides a very quick means of assessing the quality of the cash management of a project. It highlights problems and successful projects so that management action can be taken to improve future performance. There are other possible measures proposed in the literature.

Interest based methods

Ashley and Teicholz (1977) proposed the measurement of cash flow management performance, using the interest earning capacity of the net cash flow. They concentrated on the total amount of interest and the net present value of the net cash flow stream. It was in the latter context that they developed their unbalanced bidding model, on the basis that altering the timing of the cash flow stream would produce more favourable performance.

Optimisation with limited resources

An alternative method is provided by Elazouni and Metwally (2000) who constructed software based on schedules and sensitivity analysis to investigate the financial impact of resource decisions. They assumed an overdraft interest rate to calculate net cash flow plus interest cost (examination of their modelled profiles indicates that their model generates consistent profiles which follow a traditional path including substantial negative net cash flow). Their framework is as follows:

The net cash flow at the end of the month (t = 0, 1, 2..., n), excluding interest expense which is consistent with net cash flow as defined here, is:

$$A_t = P_t - E_t \tag{6.22}$$

where E_t is the construction expense and P_t is the income from the client. The cumulative cash flow at the end the month t before receiving payment P_t is:

$$F_t = N_{t-1} - E_t \tag{6.23}$$

where N_{t-1} is the cumulative net cash flow to the previous month $t - 1$. Therefore, the cumulative cash flow at the end the month t after receiving payment P_t is:

$$N_t = F_t + P_t = N_{t-1} + A_t \tag{6.24}$$

The monthly interest accumulated on the overdraft at borrowing rate i for month $t - 1$ becomes:

$$\hat{I} = I_t + \tilde{I}_t \text{ where } I_t = i N_{t-1} \tag{6.25}$$

and the approximate interest on E_t is:

$$\tilde{I}_t \cong i \frac{E_t}{2} \tag{6.26}$$

If payment of interest is deferred (added to the net cash flow) then Elazouni and Metwally (2000) argue that additional overdraft is required, giving a cumulative cash flow F_t immediately prior to receiving payment P_t of:

$$\hat{F}_t = F_t + \sum_{k=1}^{t} \hat{I}_k (1+i)^{t-k} \tag{6.27}$$

with the cumulative cash flow after payment P_t is:

$$\hat{N}_t = \hat{F}_t + P_t \tag{6.28}$$

$$\text{and } \hat{G} = \hat{N}_n \tag{6.29}$$

where $\hat{G}$ is the gross operating profit less financing cost for a project.

This approach is limited by its adoption of an overdraft rate, and a dual interest rate approach may be more suitable to modern construction management.

Total interest earned

The net cash flow was expressed by Ashley and Teicholz (1977, their equation 5) as:

$$N_t = P_t - C_t \quad (6.30)$$

where N is the periodic net cash flow for month t, P is the total payment for that month and C is the total expenditure for month t. The cumulative or total net cash F for the project to time t is given by:

$$F_t = \sum_{j=0}^{t} N_j \quad (6.31)$$

Ashley and Teicholz used a cost of finance rate r to calculate the interest cost of a net cost:

$$I_t = F_t\left(\frac{r}{100}\right); F_t < 0 \quad (6.32)$$

and a cost of capital rate s to calculate the benefit of interest earned for the reverse context:

$$Q_t = F_t\left(1 + \frac{s}{100}\right) \quad (6.33)$$

It is unclear why the different approaches were used, and the assumption that an interest cost should be used compared with the benefit cost of capital is challengeable. In reality projects do not stand in isolation and the organisation must consider the context of a portfolio of projects (Chapter 7) which implies that a single rate is satisfactory. This assumption in turn would fail in the context of a company deliberately operating projects with an overdraft facility.

The total interest earned to time t would be:

$$I_t = \sum_{j=0}^{t}\left(F_j \frac{s}{100}\right) \quad (6.34)$$

The problem with this approach is that it works on discrete monthly totals. In reality, the amount of interest would vary during the month, but this is a reasonable approximation.

Net Present Value

Ashley and Teicholz (1977) also proposed the use of the Net Present Value as an indicator. This was also proposed by Kenley (1987) and tested by McElhinney (1988). Ashley and Teicholz gave the following calculations (his Equations 9 to 11):

$$W_t = \frac{I_t}{\left(1+\frac{s}{100}\right)^t}; F_t < 0 \tag{6.35}$$

$$W_t = \frac{Q_t}{\left(1+\frac{s}{100}\right)^t}; F_t \geq 0 \tag{6.36}$$

$$W = \sum_{i=1}^{T} W_t \tag{6.37}$$

in which W_t is the month's t contribution to the net present value and W is the total net present value for a total T months.

McElhinney (1988) suggested the weighted average cost of capital to be the appropriate rate. From a small sample of eight projects, he identified clear evidence that projects can achieve their desired margins (the usual measure of profit) but be managed so poorly in terms of the timing of payments and receipts, that this margin is eroded by the cost of interest. Undertaken on data from a period of high interest rates (the mid 1980s), McElhinney's projects were greatly affected by the cost of capital at the time. The contribution of interest to the project varied from – 4.07% to 4.2% of project value, which was greater than the target margin for the projects at the time.

Internal Rate of Return

The Internal Rate of Return is a measure of the effective interest rate earned on the initial investment required for a project. However, the initial investment for a project may be small, or non-existent, and thus the internal rate could be unrealistic, and comparisons made using this measure would be misleading. For this reason McElhinney (1988) considered the Net Present Value to be a better indicator than the Internal Rate of Return, as an efficiency measure for the net cash flow management of a project. He observed that even though dependent on the selection of an appropriate discount rate, this is no different to property analysis and would likely result in measures of performance relative to the discount rate and thus comparable. The advantage of the Internal Rate of Return is that such a measure does not require the application of discount rates, and thus would be independent of the financial environment, allowing comparisons to be made between companies.

CONCLUSION

Project net cash flow is an important contributor to the successful management of construction companies. Originally driven by increased interest rates in the 1980s, interest in this topic has been maintained through recognition of the opportunity cost associated with funds generated through projects. Different models have been constructed to help understand net cash flow, dominated by methods which assume a relationship between inward cash flow and outward cash flow based on a method of weighted delays.

There has been a great deal of variability identified in project data. Understanding this variability is the key to understanding the contribution which individual projects may make to the organisation as a whole.

REFERENCES

Ashley, D. B. and Teicholz, P. M. (1977). 'Pre-estimate cash flow analysis'. *Journal of the Construction Division, American Society of Civil Engineers*, Proc. Paper 13213 **103** (C03): 369–379.

Balkau, B. J. (1975). 'A financial model for public works programmes'. *National ASOR Conference*, Sydney, ASOR.

Bromilow, F. J. (1978). 'Multi-project planning and control in construction authorities'. *Building Economist* (March): 208–213.

Bromilow, F. J. and Henderson, J. A. (1977). *Procedures for Reckoning the Performance of Building Contracts*. Highett, Commonwealth Scientific and Industrial Research Organisation, Division of Building Research.

Cleaver, H. L. (1971). 'Flexible financial control in the construction industry'. *Building Technology and Management* **9**(8): 6,7&17.

Elazouni, A. M. and Metwally, F. G. (2000). 'D–SUB: Decision support system for subcontracting construction works'. *Journal of Construction Engineering and Management,* **126**(3):191–200.

Gyles, R. V. (1992). *Royal Commission into productivity in the building industry in New South Wales*, Report of the hearings Part 1, Volume 3. Sydney, State Government of New South Wales.

Hardy, J. V. (1970). Cash flow forecasting for the construction industry. PhD Thesis, Deptartment of Civil Engineering. Loughborough University of Technology, Loughborough.

Henry, J., Holt, G. D. and Harris, P. T. (2000). 'An investigation into predicting materials suppliers' profits'. *Journal of Construction Procurement* **6**(2): 231-243.

Jepson, W. B. (1969). 'Financial control of construction and reducing the element of risk'. *Contract Journal* **24**(April): 862–864.

Kaka, A. P. (1995). 'Incorporating risk into contractors cash flow forecasting and planning'. *Financial Management of Property and Construction Conference*, Newcastle, Northern Ireland. 468-478.

Kaka, A. P. (1996). 'Towards more flexible and accurate cash flow forecasting'. *Construction Management & Economics* **14**: 35–44. www.tandf.co.uk

Kaka, A. P. and Price, A. D. F. (1991). 'Net cashflow models: Are they reliable'? *Construction Management and Economics* **9**: 291–308.

Kenley, R. (1987). Construction project cash flow modelling. PhD Thesis, Department of Architecture and Building, University of Melbourne, Melbourne.

Kenley, R. and Wilson, O. D. (1986). 'A Construction Project Cash Flow model—an Idiographic Approach'. *Construction Management and Economics* **4**: 213–232.

Kenley, R. and Wilson, O. D. (1989). 'A Construction Project Net Cash Flow Model'. *Construction Management and Economics* **7**: 3–18.

Kerr, D. (1973). Cash flow forecasting. PhD Thesis, Department of Civil Engineering. Loughborough University of Technology, Loughborough.

Khosrowshahi, F. (1991). 'Simulation of expenditure patterns of construction projects'. *Construction Management and Economics* **9**(2): 113–132.

Khosrowshahi, F. (2000). 'A radical approach to risk in project management'. *16th ARCOM Conference*, Glasgow, Glasgow Caledonian University.

Mackay, I. B. (1971). Cash flow forecasting by computer. Deptartment of Civil Engineering. Loughborough, Loughborough University of Technology.
McCaffer, R. (1979). 'Cash flow forecasting'. *Quantity Surveying* (August): 22–26.
McElhinney, K. (1988). Efficiency measures of working capital management for building projects. Masters thesis, submitted in partial fulfilment. Department of Architecture and Building, University of Melbourne: pp 50.
Navon, R. and Maor, D. (1995). 'Equipment replacement and optimal size of a civil engineering fleet'. *Construction Management and Economics* **13**: 173–183.
Navon, R., (1995). 'Resource-based model for automatic cash flow forecasting'. *Construction Management and Economics* **13**: 501–510.
Nazem, S. M. (1968). 'Planning contractor's capital'. Building Technology and Management October: 256–260.
Neo, R. B. (1979). 'Erosion of profit in construction projects'. *Quantity Surveying* August: 31-32.
Peterman, G. G. (1973). 'A way to forecast cash flow'. *World Construction* (October): 17–22.
Peters, G. (1984). 'Project cash forecasting in the client organisation'. *Project Management* **2**(3): 148–152.
Punwani, A. (1997). 'A study of the growth-investment-financing nexus of the major UK construction groups'. *Construction Management and Economics* **15**: 349–361.
Trimble, E. G. (1972). 'Taking the tedium from cash flow forecasting'. *Construction News* **9** (March).
Tucker, S. N. and Rahilly, M. (1982). 'A single project cash flow model for a micro-computer'. *Building Economist* (December): 109–115.

7

Organisational cash management

INTRODUCTION

Organisational cash flow management is fundamental to any business and construction is no different, yet it is very poorly understood in the construction industry. As a direct result, companies continue to fail, often quite spectacularly, due to a fundamental breakdown in their financial systems. The literature consistently points to cash flow management as being essential for the successful management of a construction business. This statement is often made in reference to gross cash flows, always made in reference to net cash flows, and just occasionally made where it really matters, in relation to organisational cash flow management.

There is a small and emerging body of research on organisational cash flows and a growing awareness in the broader reference literature of the role of organisational cash flow management. It is a logical extension from those papers which discuss the net cash flow position of projects to draw conclusions about the impact on organisational working capital. However, the research community seems generally unable to see through the intellectual snow-screen thrown up by industry practitioners. They have taken to believing the simple explanations of an industry which is reluctant to admit to its true financial management (and sometimes lack of management) systems, and have avoided the difficult empirical task of real data collection and modelling. This is largely due to the unnecessarily emotive secrecy which cloaks this issue within the industry.

This chapter deals with the relatively straightforward issue of organisational cash flow and its contribution to the organisation's working capital. The emotive issues surrounding the use of trade credit and exploiting the cash flow stream will be explored in Chapter 8. However, it is those practices, and their notable absence in most previous discussions of organisational cash flow, which cast doubt upon the rigour of the existing research.

In this discussion of organisational cash flow, the emphasis shifts from local site management to broader head-office management. It becomes clear that the true task of a successful construction company is not, as often believed, to manage individual projects, but rather to successfully manage a portfolio of projects. Just as there is a shift from site to central management, there is a shift from task (project) management to strategic (organisational) management, from the specific to the general.

The focus of research into cash flows must make a corresponding change. There needs to be a shift from studying the cash flow at the project level, to studying the cash flow at the strategic level. There is a degree of overlap in these views of cash flow. Those researching the project cash flow can take a bottom up approach, they study the organisational cash flow through projects. Alternatively those studying the financial management and viability of construction companies can take a top down approach, and come to studying projects through company level accounts.

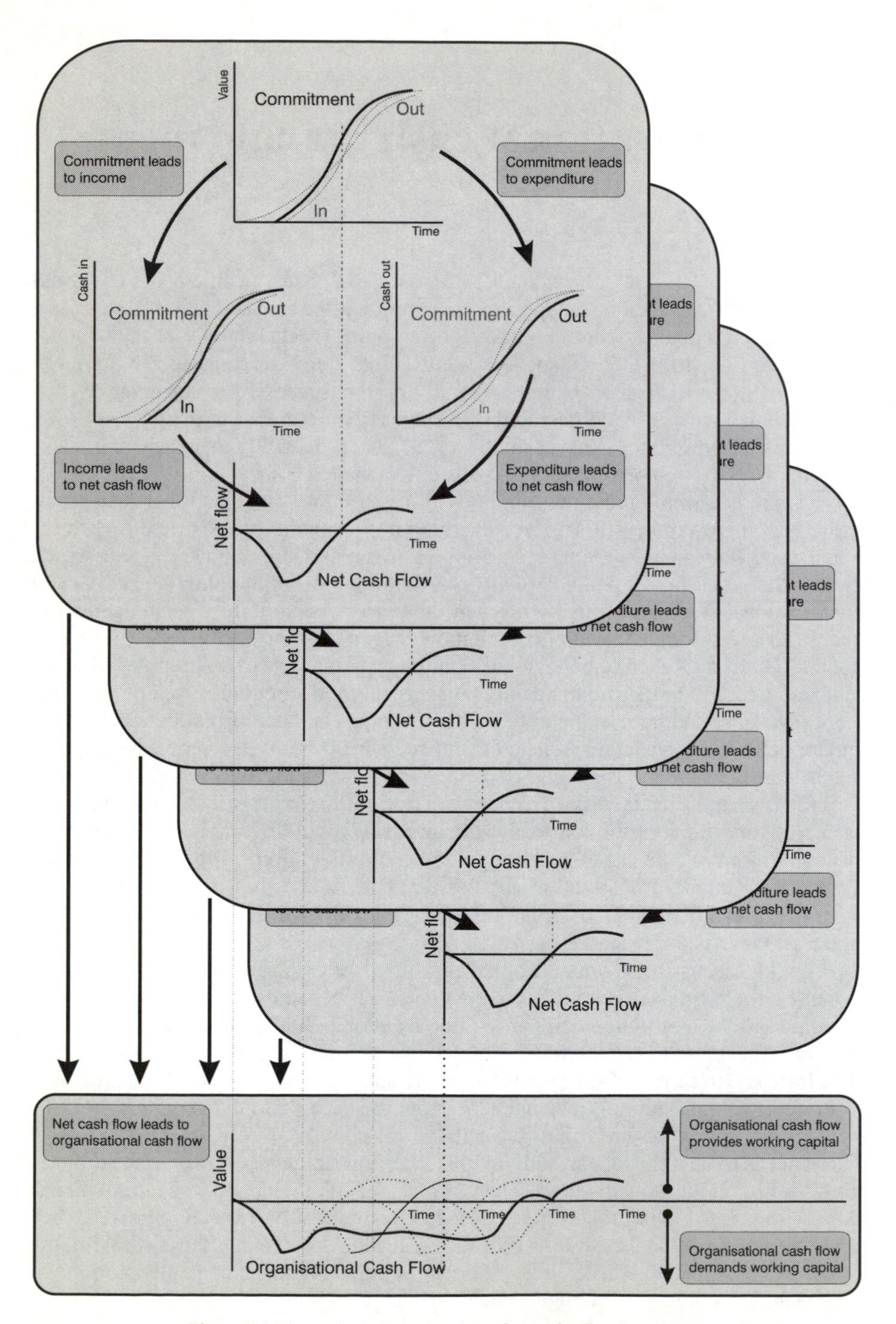

Figure 7.1 Theoretical representation of organisational cash flow

There can be no doubt of the importance of these topics to successful management of construction companies, whatever the origin of the research attention. The vast majority of sudden failures of hitherto successful companies can be traced back to a failure in their organisational cash flow management. It matters little that a profit is made on a project if the organisation cannot maintain its working capital. The drama of a shortage in liquidity must be one of the most painful and traumatic conditions for a company, never to be forgotten by those that have experienced it, whether by surviving or failing.

THE NATURE OF ORGANISATIONAL CASH FLOWS

Organisational cash flow is the next step on from net cash flow. Figure 6.3 illustrated the theoretical construction of the net cash flow. Similarly, the organisational cash flow can be seen to be constructed by a portfolio of project net cash flows. This is illustrated in Figure 7.1.

The portfolio view considers the cash flow of the organisation to be constructed from a series of overlapping individual net cash flows. Where they overlap, they accumulate, the net inward cash flow on one project countering the net outward cash flow on another. This is an important concept in traditional construction management. It is very common for the early phases of a project to require subsidy funding. This can be offset by the positive balance of another project reaching completion, thus maintaining corporate liquidity.

Policy and organisational cash flow

Organisational cash flow reflects the business culture of the organisation. A traditional, conservative firm will operate with a different form of organisational cash flow from a modern, more aggressive organisation. A company can be aware of its position and manage its cash flow in a strategic manner, or it can be unaware and run its cash flow as an incidental occurrence, being focussed solely on the project margin.

While such differentiations are an oversimplification, they highlight different policies toward managing the cash flow of a business. To explain further, it is necessary to examine again cash flow from first principles, returning to the basic concept of gross cash flow.

A project net cash flow is developed from both inward and outward gross cash flows, with the final balance being the final margin for the project. An idealistic version of this is shown in Figure 7.2, however, this profile is uncommon on traditional projects and a more realistic traditional profile is that shown in Figure 7.3, where inward payments lag outward payments. This project provides negative funds contribution in the early stages and struggles to achieve its intended final project profit margin, with break-even being well advanced through the project. The project is an overall drain on company resources.

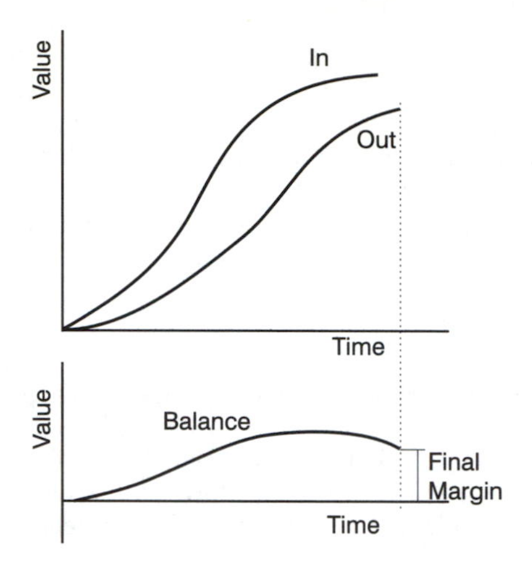

Figure 7.2 Idealised project gross and net cash flow

We are used to thinking about projects demanding resources. It follows that in order to take on a tendered project, we think about whether we have the resources to complete that project, including the financial resources.

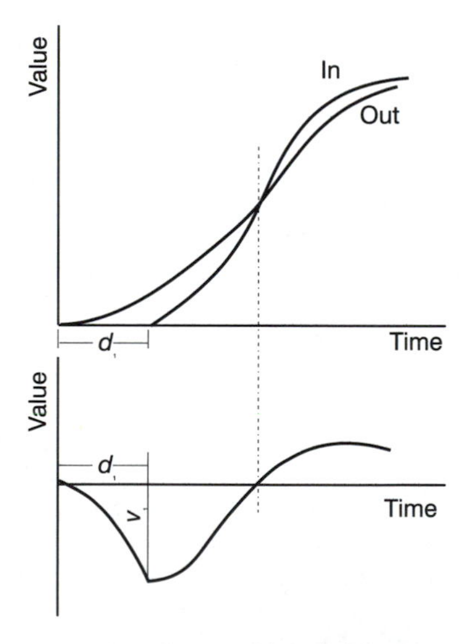

Figure 7.3 Realistic traditional project gross and net cash flow

A common approach to managing projects results in outgoings being paid before, and continuing to lead the corresponding income through most of the project . There is a lengthy start-up phase resulting in a corresponding lengthy period before break-even. Such a project has the profile indicated in Figure 7.3 and requires funding, normally overdraft funding, to the amount indicated by v_1. This places two demands on an organisation, to be able to establish an overdraft limit sufficient to cover the maximum demand and to fund the overdraft interest amount. In turn, this also places limits on the operation of the organisation, if one project absorbs the available overdraft limit, then no further projects can be undertaken until that project is completed or has at least sufficiently reduced its demands on the financial resources to allow further projects. This period of negative cash flows was described as 'capital lock-up' by McCaffer and Harris (1977) to 'represent locked up capital that is supplied from the company's cash reserves or borrowed'. In fact it is the working capital of the company which is locked up through this mechanism.

It is a common situation for small builders to be limited to one project at a time through the inability to raise further finance for a second project. The same can occur at the opposite extreme, where a successful company can reach the point where they cannot afford to win further projects until some of the existing workload is completed and the financial demands are reduced. Projects run with this form of net cash flow profile have an overall negative impact on the total organisational cash flow.

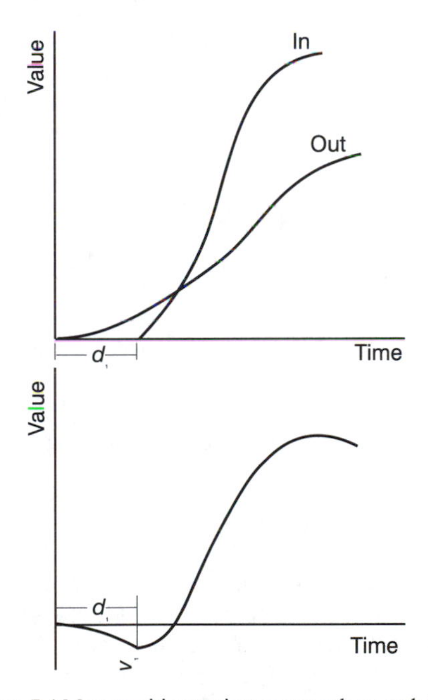

Figure 7.4 More positive project gross and net cash flow

A preferable approach to managing projects results in income either starting first (which may be difficult) or at least as soon as possible after expenditure has started, thus reducing the delay before income commences and the consequential demand which the project places on organisational working capital. Only small changes are required to drive the project in such a way that the project becomes a funds contributor, as net cash flow curve profiles are very sensitive to project gross cash flow conditions.

The profile displayed in Figure 7.4 is for a project which provides funding through most of its life. This reduces the extent of required overdraft funding, together with interest cost. In fact there is the potential to earn income from interest on funds generated. Similarly, the ability to take on additional projects is only a problem in the early stages, and ceases to be an issue after the early stage, prior to break-even.

An organisational portfolio of projects

The impact of these different approaches is seen when an organisation is running a portfolio of projects. In those circumstances the ability of the organisation to take on new projects, to fund those projects and to continue to fund its normal operations, becomes paramount.

This concept was first published by Nazem (1968) who, in his Figure 2, overlaid many project net cash flow curves, albeit with each of the same profile, but with varying scale. This important observation of Nazem may be taken further, with the effective cumulative value of the portfolio of projects being represented. This concept of overlaying individual net cash flow profiles to form an organisational cash flow profile is the key to Figure 7.1, with the overlay being detailed in Figure 7.5.

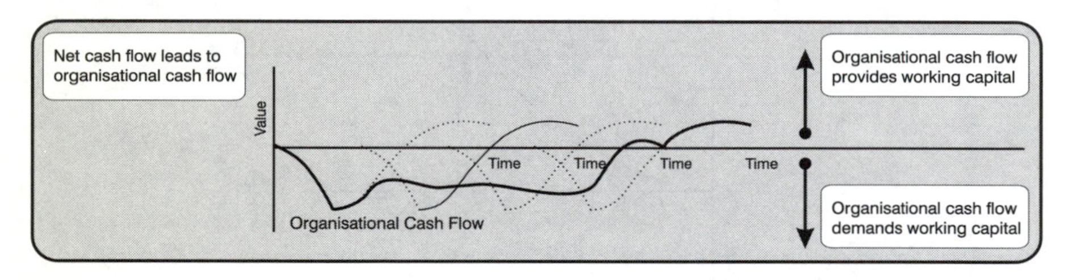

Figure 7.5 Overlaying net cash flow profiles to form an organisational cash flow

Standard or variable cash flow profiles

Only one stream of literature to date has considered the possibility that there is a range of possible net cash flow profiles when considering organisational cash needs. Most other literature has followed the belief that income lags outgoings in a constant fashion. Kenley (1987) argued that the net cash flow profile was unique and that the sensitivity of the net curve to the timing of payments was such that a very small change in the originating curves has a dramatic effect on the resulting net cash flow curve. This idiographic approach was also followed by McElhinney (1988) but the belief on which it was founded has since been largely ignored. This is surprising because the empirical data does not support the alternative—to construct net cash flows from standard delays. There is a gap between the dominant theory and the observable reality.

The reality is that there are mechanisms at work which will not be admitted to on a researcher's survey form, which are not part of the taught construction management curricula, but which dominate the policy and practice of construction firms.

There are many questionable methods by which a contractor may manipulate the timing of payments, including front-end loading rates (although this dubious practice has been accepted by the research community as normal—most likely because it was recognised by Ashley and Teicholz (1977) and faithfully repeated thereafter) and delaying payments. However, a contractor may also achieve this end by entirely legitimate means, managing the timing of payments in much the same way as a bank or other financial institution manages its cash flow. The most obvious method is to carefully manage the timing of payments through contract terms. There has been very little research into the true meaning of terms of payment and at this stage it is useful to divert for a discussion about standard conditions of payment and the meaning of contract terms.

CONTRACT PAYMENT TERMS—WHAT DO THEY REALLY MEAN?

Models of cumulative multi-project organisational cash flow have assumed that estimated receipt–disbursement time–lag factors can be used to calculate monthly cash flow figures, and that these once calculated can be overlaid cumulatively. This approach is supported by Archer (1966) who analysed the literature and concluded that monthly transaction or activity summaries could be summed to arrive at what was termed a 'stock of cash' needed for the management of the firm. While there is agreement that overlaying individual project net cash flows will result in the net organisational contribution, there is no such agreement on the method for modelling the net cash flow source curves.

If the assumption that a standard delay may be applied is to be utilised, then it is necessary to increase understanding of the types and extent of delay involved in the management of cash flow by construction companies. These delays relate back to the work in progress curve and apply to both the flows of cash in from the client, and the flows of cash out to the sub contractors.

Inward cash flow delays relate to the delay between the work being completed on the site, and the payment by the client. This delay relates to the delay in claiming (typically at the end of the month) and the subsequent delay while the client values the work and then processes the payment and related adjustments for retention. These delays may be assumed to relate to the contractual terms applicable and could vary between one week from the end of the month in which the work was carried out, to one month from certification which might be two weeks from the end of the month in which the work was undertaken (a total of approximately six weeks). Individual circumstances, client and certification team practices, contracts, etc. drive these delays.

In contrast, outward cash flow delays relate to the financial policy decisions taken by individual contractors. Rather than define weighted average delays based on individual projects, individual contractors will have delays which apply based on their own financial management policy, including the supplier/subcontractor payment system.

Accordingly it is necessary to examine individual contractor companies to assess their cash flow management policies and the implications which this might have on a delay based analysis; and which these might, in turn, have on the overall financial management of the company.

An analysis of contract terms

Contractors have different interpretations and therefore management of their trade credit, which has significant implications for organisational working capital. In order to demonstrate this belief, a pilot study was established to analyse individual contractor's understanding of cash flow management and their policies regarding payments. This work was first reported in Kenley and Wood (1996), and remains an area requiring urgent further investigation.

Method

A study of eleven contractors was conducted, with the objective of determining where future research in the area of contractor credit control could be directed. The contractor sample was chosen to include medium to large employers of subcontractors, from both country and metropolitan Victoria, Australia. The method of data collection was by way of a structured interview, employing both closed and descriptive questions.

The questions sought to investigate the following financial policy issues which relate to cash flow management:

- The management of retention.
- The management of the timing of payments.
- Adherence to contractual conditions.

Contractor's were asked questions about the management of their individual projects and to provide responses by the relevant percentage of their turnover that would provide an affirmative response.

Results

To assess the management of retention, contractors were requested to provide their average annual turnover for the past three financial years and whether it was company policy to allow subcontractors to provide bank guarantees in lieu of cash retention and, proportionally by value, what percentage of subcontractors' work would be included in each category. They were also requested to indicate their policy toward the provision of security to the client.

When asked whether it was company policy to provide bank guarantees to the client when allowed under the contract, respondents overall indicated 80% of turnover was under bank guarantee rather than retention. The results in Table 7.1 show that in contrast, although approximately 60% of subcontractors' work is subject to cash retention by contractors, but that this equates to a weighted average of only 42% by value. On average 75% by value indicated interest earned on that subcontractor cash retention was not returned to subcontractors.

Table 7.1 Percentage of subcontractor's work, in cash or bank guarantee by turnover (adapted from Kenley and Wood, 1996)

subcontract agreements	1	2	3	4	5	6	7	8	9	10	11	Avg	Wtd avg
Cash retention %	70	60	50	50	0	60	50	70	0	0	30	40%	58%
Bank guarantee %	30	40	50	50	100	40	50	30	100	100	70	60%	42%
Turnover $m	300	75	30	300	30	30	300	300	4	8	2	125	

The general policy then appears to be, as expected, that contractors utilise bank guarantees to avoid cash retention being held by clients, but receive both cash retention from subcontractors and consequently interest payments. Thus, considering weighted average delays only, and not referring to financial management policies, would give an erroneous financial picture of overall trade credit management. Consider, that by mid-contract on a ten million dollar contract at 5% subcontractor retention, the contractor may have at its disposal $500,000 plus interest accruing. By disregarding this cash input which has no corresponding cash outlay (as the contractor has provided bank guarantees) the true financial position of a company would be distorted.

There is, of course, no assurance that the cash retention/bank guarantee scenario would be the same on every project. This finding is supported by the pilot survey (refer to Table 7.1) and is consistent with the Kenley and Wilson (1989) case for non-standard delays. Therefore, unique projects may have differing cash retention/bank guarantee policies. Any model developed must, once applied, account for differing contract conditions, trade emphasis and client certification procedures.

In order to understand the management of trade credit, or terms of payment, contractors were asked for their standard terms of payment for different work trades. In order to ensure that terms have equivalency, they were also asked to describe what they understood 'thirty-day terms' to mean in an attempt to identify common interpretation of the term 'thirty-day payment'. Understanding the definition and use of such terms is critical when developing any model which addresses the management of trade credit, such as weighted mean delays.

The construction industry has overwhelmingly adopted the general credit term of 'thirty days'. When asked for their general terms of payment for subcontractors and suppliers, the answers were unanimous in stating thirty days (although minor exceptions were made by most for labour-only subcontracts particularly formworker, reinforcement fixer, bricklayer and carpenter). However, when asked what they understood thirty-day terms to mean, the answers were diverse. Of the ten companies who responded to this question, the following definitions were provided:

1 Payment thirty days from valuation of certification of work completed.
2 Payment thirty days from invoice date.
3. Supplier's/subcontractor's claims in by the 25th of the month, payment on the 30th of the following month.
4 Supplier's/subcontractor's claims at end of the month, payment thirty days later
5 Payment thirty days from date of invoice.

6 30th day of month following receipt of invoice.
7 Payment within thirty days of invoice being presented.
8 Payment thirty days from valuation—back to back, i.e. client pays thirty days after head contract claim, suppliers/subcontractors paid after head contract payment.
9 Payment thirty days from end of month that claim presented.
10 Payment made by the end of the next month goods/services invoiced.

It is interesting to compare these individual definitions (all provided by the finance director or managing director) with each other and with accepted accounting definitions. First, there is surprisingly little repetition in interpretation. Secondly, almost all of the definitions adopted by the building firms examined are beyond the accounting definition provided by Warren and Fess (1989), who defined 'net thirty-day terms' to mean thirty days from the date of invoice.

The replies indicate that any model which uses delays based on a single definition of 'thirty-day terms' should be treated warily. Thirty days takes on various meanings according to the above definitions Therefore, if we take a scenario which has work completed and accepted on June 5 and invoice dated June 12 but presented two days later, Table 7.2 gives the various ranges for the term 'thirty-day payment' resulting from the responses above. Also presented are the equivalent delay figures for the instance when the invoice is not presented until July 1.

Table 7.2 Delay from completion of work to payment (Kenley and Wood, 1996)

Respondent (contractor)	Date of payment (to supplier/ subcontractor)	Actual delay with invoice received 14/6	Actual delay with invoice received 1/7
1	5 July	30	30
2	12 July	37	37
3	31 July	56	86
4	31 July	56	86
5	12 July	37	37
6	31 July	56	86
7	3 July	27	47
8	7 July	31	86
9	31 July	56	86
10	31 July	56	56
Accounting std	12 July	37	37

The variance in definition of 'thirty-day payment' results in a range of the actual number of days from work completed to payment of 27–56 days. The comparison with the accounting definition can be viewed in Figure 7.6.

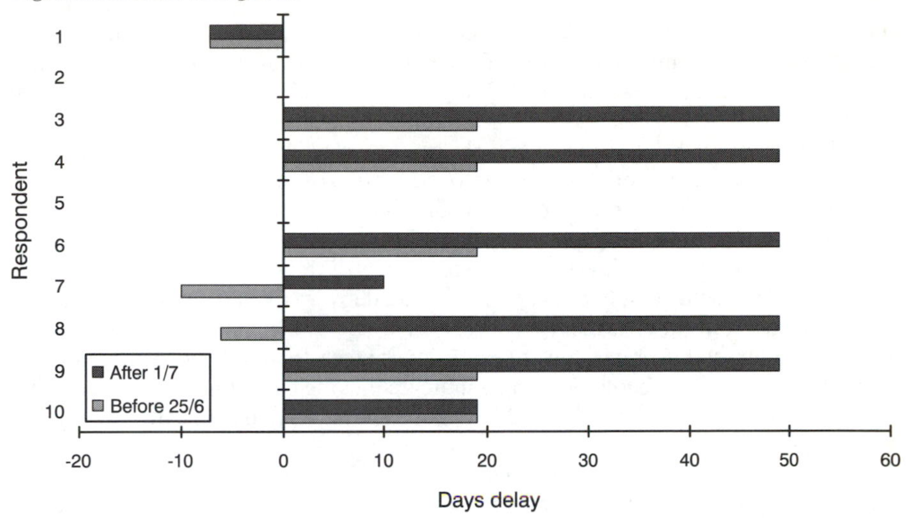

Figure 7.6 Chart of payment delay compared with the accounting definition (Kenley and Wood, 1996; Kenley, 1999)

It is clear that using the accounting definition of 'net thirty-day payment' and applying weighted delays without taking into account the actual company policy would not be representative of the industry. Further to this it can be seen that the situation can be exaggerated, particularly when receipt of an invoice is delayed.

The supplier who is three days later than the cut-off deadline with forwarding their invoice, can actually be paid ninety days after completing their work. If we take, for example, respondents 3 and 9 definition of thirty days, a claim made three days after the 25th (or the end of the month) would result in a payment being made to those suppliers on 30th August or a period of 86 days after work complete. If the contractor had claimed from the client at the end of June, as would most probably be the actual situation, and therefore received payment at the end of July, then the weighted delay for the contractor in respect of their cash flow would be recorded as positive; again with accruing interest payments.

A premise of much research to date has been that contractors strictly follow the contract terms with regard to payments. It is assumed payments to suppliers and subcontractors are made on dates specified in the contract: usually claimed by the 25th of the month with payment made on the 30th day of following month. However, the results of the pilot study indicate that payments are frequently withheld as a carrot and stick method of controlling suppliers and subcontractors. Asked whether payments to suppliers are ever **delayed** for other than contractual reasons, for example slow progress, five of eleven replied yes. Similarly, when asked whether subcontractor payments were ever **withheld** for reasons other than contractual, i.e. as a penalty for being behind programme, disagreed back-charges, etc., eight replied yes.

Thus company policy could have potentially significant implications on the company operational cash flow, especially if it is found to be consistent throughout the course of the contract. It is clear that contractor financial policy may have hitherto unknown implications on the development of mathematical models for analysis of the net cash flow on any given project. This in turn impacts on any overlaying of individual project net cash results required to give an overall company working capital profile.

The accounting definition of 'net thirty-day terms' being payment thirty days from date of invoice was only being employed by two of the eleven contractors. The remainder have a subcontractor payment distribution range between twenty-three and forty-nine days (and longer in certain circumstances). This would obviously impact on any weighted delay analysis of project cash flow. Furthermore, the reason for change in the delay is subcontractor error (late delivery of claim) and therefore not predictable by conventional means.

The contractor's policy toward bank guarantees for the head contractor and cash retention for the subcontractor may indicate that the cash flow may be receiving a considerable boost, and weighting delays could be applied in the positive as well as the negative.

Respondents indicated that it was common for contractual obligations to be overlooked if the withholding of payment served as an incentive for the subcontractor to increase rate of work. Again, if the suspension of subcontractor payments is in any way prevalent, this could have a marked effect on project and organisational cash flow.

The significance of these results is that previously held beliefs of conformity in the application of payment terms, retention policy, and adherence to contract conditions have been brought into question. The sample is not large enough to be statistically significant and therefore does not establish a complete picture of contract terms; however, it is sufficient to show that there is a problem in the interpretation and analysis of existing arrangements. If the wide variance from accepted norms is, in fact normal, then models of working capital derived from building projects may need to be reviewed. Future research must make allowance for this variation in the mathematical models developed.

INVESTIGATING ORGANISATIONAL CASH FLOW

Introduction

Past research into the causes of bankruptcy in the building industry has been dominated by case study research, looking at failed companies in order to identify likely causes of failure. Following one such study, Kaka and Price concluded that

> ...the construction sector usually experiences a proportionally greater number of bankruptcies than other industries...the final causes of bankruptcy are inadequate cash resources and failure to convince creditors and possible lenders that the inadequacy is only temporary (Kaka and Price, 1991a).

That inadequate cash resources are a cause of failure would rarely be disputed. However, the conclusion that temporary shortages of working capital may lead to failure might be challenged. The shortage of working capital may be real, and the shortage temporary, but the shortage is predictable and normal. Therefore, the fault is one of management rather than an inability to convince creditors.

Building operations once used to be funded by the builder, typically with an overdraft or less frequently with accrued capital. Increasingly this practice has shifted to an alternative approach, whereby firms use trade credit and early payment from the client to cover (and sometimes exceed) the costs of operations. It has been shown by Nicholas (2000) and Punwani (1997) that trade credit is a significant component of the financial structure of the industry, this places a large burden on the subcontractors and suppliers who unwittingly provide it, sufficiently so that it directly impacts on their ability to conduct their business.

Management of the contractors working capital is the essential issue. It seems reasonable to expect that a significant factor in bankruptcy is poor cash management, as stated by Paté-Cornel *et al.* (1990) '...cash–flow remains the ultimate test of survivability'. However, does it follow that firms which do not fail, have managed their cash flow, and from it their working capital, well? Moreover, what does this mean for a successful company?

The next step in the exploration of cash flow is to explore the organisational cash flow profile, to construct a simple model of a contracting organisation and to use it to simulate organisational cash flow. A stochastic model is used to simulate the levels of working capital available within a construction company that is using a specific set of cash management techniques on a large and systematic scale. The model demonstrates that significant amounts of working capital are generated, and this alone may be sufficient reason to engage in construction activity.

Method

The method uses a computer model of an organisation to calculate the levels of working capital available through building operations.

The simulation is based on a set of thirty-six hypothetical projects. These are commenced over a three-year period and modelled over a total of five years. This period was chosen to enable a stable twelve-month period to be analysed in the third year. The projects are of varying duration, and duration is correlated with the contract sum. Payment profiles are generated for each project and net cash flow is generated using an applied set of delays from assumptions about retention and delays in payments. The cumulative net cash flow for all the projects is then calculated over time. As any single set of results is meaningless, the model is based on stochastic estimating using the Monte-Carlo[1] simulation method.

In this study, stochastic estimating was applied to a wide range of the assumptions used in the simulation of working capital for the organisation.

The components of the model may be summarised as follows;
for each of the projects, the components:

- Generate project value and starting date;
- Calculate project duration from project value;

[1] The study uses the *Crystal Ball* software package from Imagineering.

- Generate cash payment profiles;
- Calculate retention applicable; and
- Apply delays and calculate funds generated for the project.

then for the twelve-month period, for the whole organisation, the components:

- Calculate turnover;
- Calculate the contribution of the projects to working capital;
- Calculate the minimum and maximum contribution provided over the select twelve-month period; and
- Calculate the minimum contribution as a percentage of turnover.

The assumptions and method of applying the stochastic estimates, including probability distributions, are outlined in the following sections.

Selection of the period for analysis

The model simulates a successful on-going concern. In order to achieve a situation where the model is simulating an organisation which has been operating successfully (rather than an organisation in a start-up phase) the model allows two years of start-up prior to the sampling period in the third year. This choice is partly due to the scale and, the duration of the projects being modelled. The model includes the final two years for observation purposes only. The twelve-month selection period is, therefore, the mid-year of a five-year analysis.

Generating project starting date

Project start date (simulating a successful bid, negotiation or some other mechanism for gaining a project) is drawn from a uniform probability distribution (Figure 7.7). The commencement date is given as a base date plus a time interval of between zero and 1065 days (35 months).

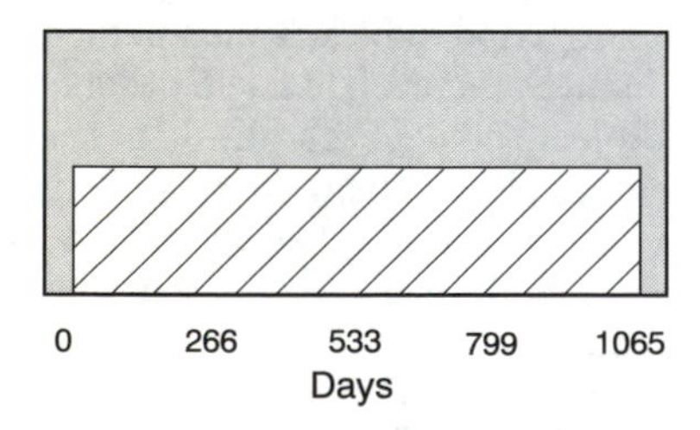

Figure 7.7 Probability distributions for start time (Kenley, 1999)

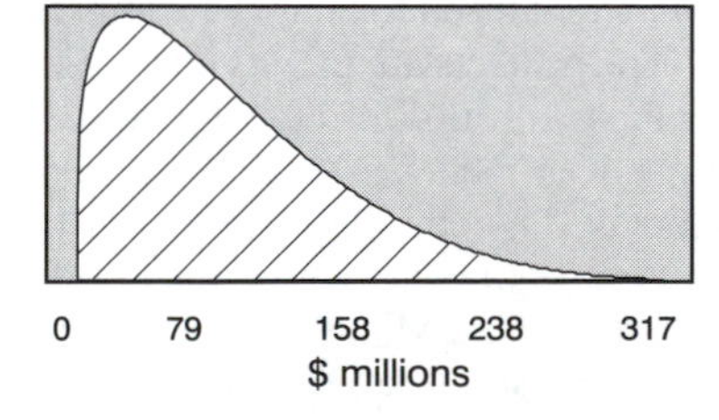

Figure 7.8 Probability distributions for project value (Kenley 1999)

Generating project value

Project value is established at the time of tender. It was assumed for the analysis that the value of a successful tender at any time has a beta probability distribution. The beta distribution is skewed (Figure 7.8) allowing a greater probability of a lower value project and relatively lower probability of a high valued project—but with a long tail. For this analysis it was assumed that the minimum tender would be $1 million, and the maximum $400 million. The beta distribution has an alpha of 1.3 and a beta value of 5. Accordingly, a project value over $300 million is very unlikely. The median is approximately $50 million.

It should be noted that project margin has been ignored in order to concentrate on the other effects of cash management without the effect of the normal business practice of charging a margin. This generates an conservative result.

Calculating project duration from project value

Project duration is calculated in the model as a function of the project value. The greater the project value, the larger the duration. The formula used for time as a function of cost is given by the equation (Bromilow and Henderson, 1977) which is:

$$T = KC^{B} \qquad \text{(5.1 repeated)}$$

where K is an efficiency variable, C is the project value in millions adjusted to September 1972 figures and B is a constant, approximating 0.3.

The variable K is lower for projects that are more efficient; Bromilow and Henderson found it ranged between 173 for a fast project and 407 for a very slow project (at the 95% level of confidence). Their average for K was 313 in 1972. There have been a number of analyses of the relationship between cost and time, which have utilised the Bromilow and Henderson equation. Analysis of these (for the complete analysis, refer to Chapter 5) indicated that there has not been significant variation over time. Table 7.3 shows some of the few values which have been provided with confidence limits.

Table 7.3 Brief comparison of estimates of B & K

Analysis	B	σ of B	K	σ of K
Bromilow and Henderson (1977)	0.3		313	85 to 93
Kaka and Price (1991b)	0.287	0.15	391	102.5

A normal distribution for variable K is assumed (Figure 7.9) having a standard deviation of approximately 90 about the mean of 313 in 1972. This value is then adjusted to current dollars using the Building Cost Index (BCI). Bromilow's model required adjustment of the project value into September 1972 equivalent dollars. While not expressed monthly in 1972, this index can be interpolated to have a value of 142 as at September 1972. The index (base 100 in 1962) remains current today, although its value was taken to be 770 at the time of the analysis.

The value for B was the approximation of 0.30 varied according to the standard deviation results identified by Kaka and Price (1991b), giving a standard deviation of 0.15 about a mean of 0.3 assumed for B. The probability distribution is shown in Figure 7.10.

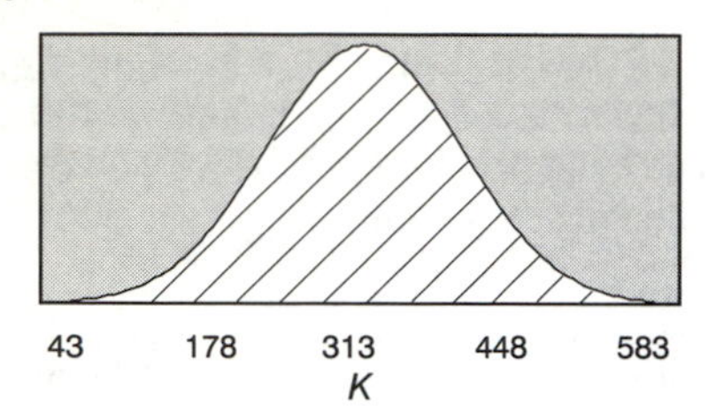

Figure 7.9 Probability distribution for time factor K (Kenley 1999)

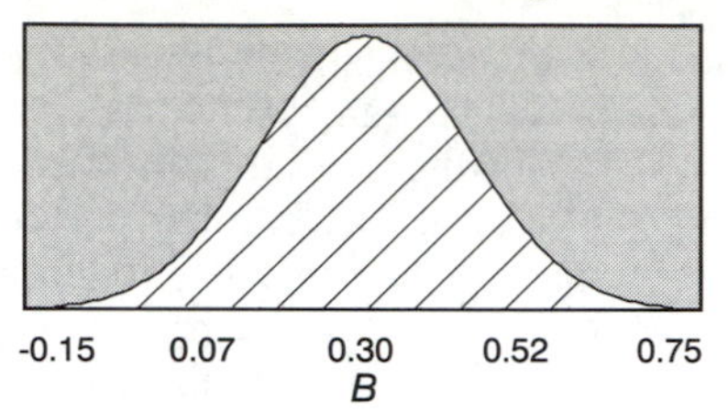

Figure 7.10 Probability distributions for performance factor B (Kenley 1999)

For this analysis the equation for time as a function of value is therefore:

$$T = K(rC)^{B} \tag{7.1}$$

where K is the time factor with mean 313 and standard deviation 90, r is the ratio calculated from the index, for this example using the adjustment to January 1996 from September 1972 (= 0.1844), C is the value of the project in current dollar terms (millions); and B is the performance factor with mean 0.3 (Bromilow and Henderson, 1977) and standard deviation 0.15.

Variation of the base model would involve considering the time performance factor, the degree of variation and the adjustment to current day values by either using an adjusted K value, or by adjusting the value of r from Equation 7.1.

Generating cash flow profiles

The gross inward cash flow from the client has been modelled by use of the Kenley and Wilson (1986) Logit transformation model (Equation 3.6).

$$v = \frac{100.F}{1+F} \quad \text{where } F = e^{a}\left(\frac{t}{100-t}\right)^{\beta} \qquad \text{(3.6 repeated)}$$

where v is the percentage of value complete, t is the percentage of time complete, and α and β are constants.

Table 7.4 Mean and Standard deviation for α and β (Kenley, 1986)

	Mean	Standard Deviation
Alpha	0.2156	0.7365
Beta	1.6421	0.4732

This model allows the generation of a cash flow profile for a range of projects by using only the two parameters α and β, and has been subject to rigorous testing since its development (see Chapter 3). An analysis of historic data from Kenley (1987) indicates a range of possible values for α and β. These values are summarised in Table 7.4, and the resultant probability distributions for α and β are illustrated in Figures 7.11 and 7.12 respectively.

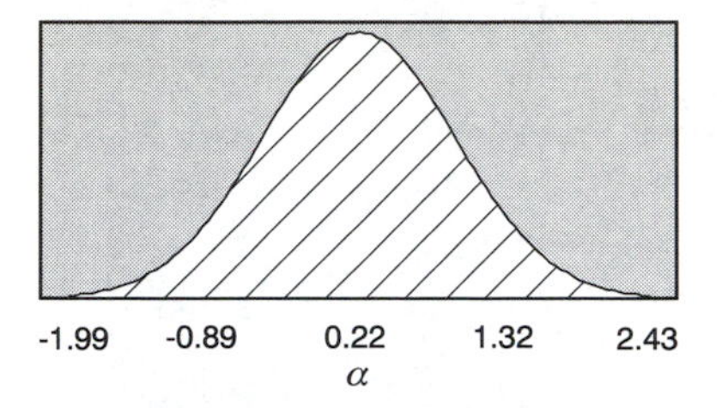

Figure 7.11 Probability distribution for parameter α (Kenley, 1999)

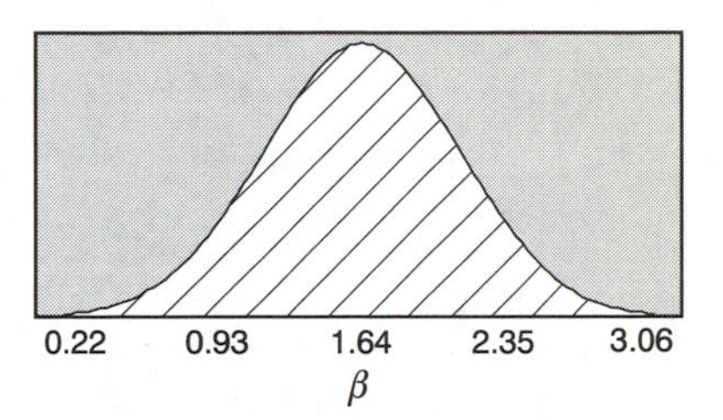

Figure 7.12 Probability distribution for parameter β (Kenley, 1999)

Further analysis indicated that α and β were positively correlated with a coefficient of 0.48. The scatter plot for α versus β and the associated line of best fit are shown in Figure 7.13. Thus, for the simulation, the values of β were positively correlated with α.

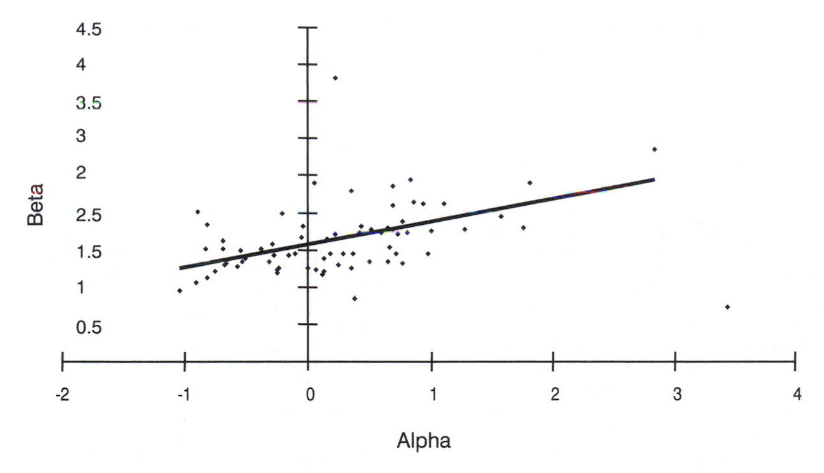

Figure 7.13 Correlation of α and β from 69 projects (Kenley, 1999)

A cash flow profile for each project can be derived from the project value, the project duration and the constants α and β. Given also the project start date, it is possible to develop monthly cumulative work in progress profiles for each project.

Figure 7.14 illustrates one possible outcome of the thirty-six individual projects', cumulative work in progress profiles over the five years analysed. It should be noted that individual iterations through the simulation would result in a completely different sample of cash flows. The unshaded area indicates the selected analysis period.

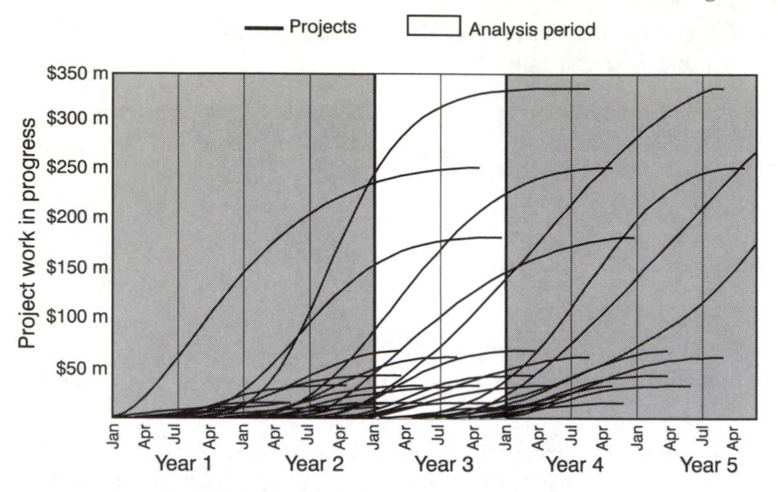

Figure 7.14 Simulated projects (Kenley, 1999)

The sum total of the work in progress for the organisation is shown both periodically and cumulatively in Figure 7.15.

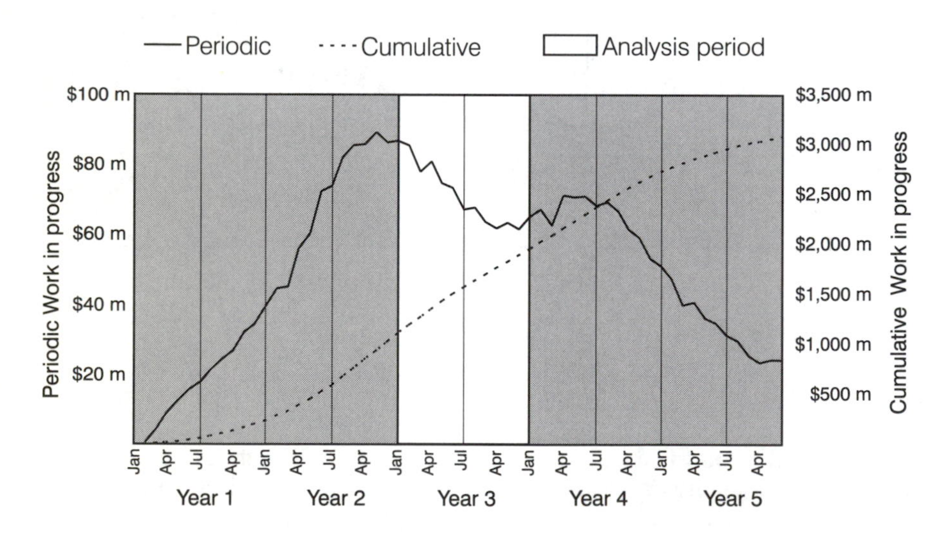

Figure 7.15 Simulated organisational cash flow (Kenley, 1999)

Calculating net cash flow for each project

Modelling the net cash flow may be done by either the method of applied systematic delays (also called weighted mean delays) or by examining the individual source curves. Kaka used an applied delay, as given in Table 7.5.

Table 7.5 Time lags for each cost category (Kaka and Price, 1991a)

Cost type	Number of months delay 0	1	2	3	Percentage of cost
Labour	100%	0%	0%	0%	15%
Plant	0%	20%	60%	20%	5%
Sub-contractor	0%	50%	50%	0%	45%
Materials	20%	70%	10%	0%	25%
Site overheads	100%	0%	0%	0%	10%
Total cost of delay	30%	41%	28%	1%	100%

Chapter 6 discussed the method of weighted mean delays. Although this method has critics, it does have merit in examining a firm's working capital policy and its relationship to the net cash flow position of the firm, particularly over a range of projects. The method has been adopted for this analysis, but it should be recognised that much greater detail is required and much more research before this approach can be considered reliable.

The method of applied systematic delays was used in the simulation to adjust the originating cash flow curve to represent the lag between outward cash flow and inward cash flow. The incoming cash flow after retention was assumed to be held after receipt and prior to payment. Table 7.6 indicates the weighting of monthly holdings used in the simulation.

Table 7.6 Holdings per month

Period	% held
Month 1	85%
Month 2	75%
Month 3	30%

These delays may be contrasted (Table 7.7) with the figures used by Kaka and Price from Table 7.5. This study uses only slightly greater delays than those estimated by Kaka and Price, and are designed to reflect an organisation systematically using the contractual terms to delay payment as long as possible in accordance with Table 7.2.

Table 7.7 Comparison of the delay amounts from Tables 7.5 and 7.6

Period	Kaka and Price	Model
Month 1	70%	85%
Month 2	29%	75%
Month 3	1%	30%

Ashley and Teicholz (1977) introduced an unbalancing technique to handle front-end loading, and this was also included in the Kaka and Price model. Techniques such as front-end loading, forward claiming and withholding variations have been ignored in this analysis, in order to test the model without such distortions. Accordingly, the assumptions in the model may be considered conservative by some standards. Future work should introduce the unbalancing method and adjustments for other factors such as materials on site (Kaka and Price, 1991a) to further refine the model.

Calculating retention

It was assumed for the analysis that the work in progress and the value of progress claims are equivalent. Kaka uses the former as a predictor because he believed them to be more reliable, however for this analysis it is the lag between the two that matters and this is taken into account when calculating applied systematic delays. The delay was assumed to be minimal. Retention was therefore calculated against the work in progress profile for each project. The assumptions applied were that:

- Retention was held at the rate of 10% up to the value of 5% of the total contract value.
- At practical completion, which for this analysis was assumed to be at the end of the project, the retention rate was reduced to 2.5%.
- At the end of the defects period (12 months), the retention was assumed to be nil.

The model is of cash flow for the contractor. It is assumed that the contractor would prefer bank guarantees in lieu of retained payments when dealing with the client, but the reverse when dealing with subcontractors. Cash retention against subcontractors would be held at 70% of the value of the work in progress, with the balance being an approximation for direct cost items where no retention is appropriate 20%, for profit margin 5%, and for preliminaries 5%, totalling 30%.

Funds generated

The total funds generated equals the total of the retention amount and the amount held for each of month 1, month 2 and month 3. Profit is retained at 5% at the end of each project.

Simulation

The simulation was run with 5000 iterations. This is equivalent to the company being formed 5000 times and each time running 36 projects of varying turnover.

Results

Within the limitations of the assumptions built into this model, the working capital available has a range over the sample year from a minimum figure to a maximum figure. The minimum figure may be seen as the safe amount available for reinvestment without significant risk.

The results support the contention that organisational cash flow, even without the benefit of front-end loading, has the potential to generate significant working capital. Table 7.8 shows the stochastic analysis for average funds generated during the subject period. This shows that the company would enjoy a significant positive contribution to organisational working capital through operations.

Table 7.8 Forecast: Average funds contributed to working capital

Statistic	Value
Mean	$200,214,387
Standard deviation	$39,158,687
Skewness	0.32
Kurtosis	3.25
Coefficient of variability	0.20
Minimum value at 95% confidence	$128,556,398
Maximum value at 95% confidence	$281,666,667

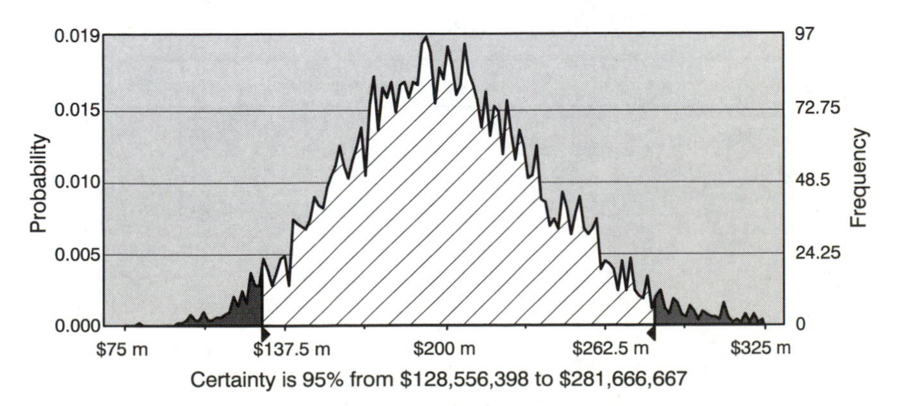

Figure 7.16 Forecast: Average funds generated

The frequency diagram which resulted from this analysis is shown in Figure 7.16 and demonstrates an approximately normal distribution of the average level of funds generated for the company during the subject period.

GENERATING FUNDS—CONTINUING THE MODEL

Table 7.9 and Figure 7.17 presents the results of the minimum funds generated for working capital during the subject period tested at the 95% confidence level.

The turnover for the twelve-month subject period is illustrated in Figure 7.18. Turnover is shown at 95% confidence between approximately $600 million and $1.4 billion with a mean approaching $1 billion. These figures are consistent with the target turnover for the model company.

Table 7.9 Forecast: Minimum funds generated for working capital

Statistic	Value
Mean	$158,981,438
Standard deviation	$35,753,348
Skewness	0.18
Kurtosis	2.97
Coefficient of variability	0.22
Minimum value at 5% confidence	$101,867,535

The minimum working capital is shown at 95% confidence of being $100 million. This approximates 10% of turnover. However, as it is possible for low minimum working capital figures to be associated with periods of low turnover, a higher ratio may be expected. Accordingly, this ratio was checked. The results are displayed in Figure 7.19 and statistics are summarised in Table 7.10.

Table 7.10 Forecast: Minimum working capital as a percentage of turnover

Statistic	Value
Mean, Median	16.04%, 16.48%
Standard deviation	3.07%
Skewness	–0.60
Kurtosis	2.94
Coefficient of variability	0.19
Minimum value at 5% confidence	10.36%
Minimum and Maximum values at 2.5% and 97.5% confidence	9.08%, 20.71%

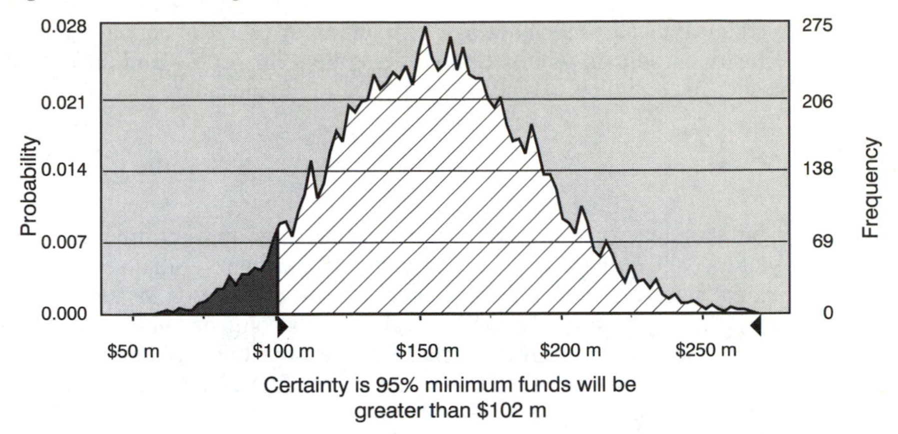

Figure 7.17 Forecast: Minimum funds generated for working capital (Kenley, 1999)

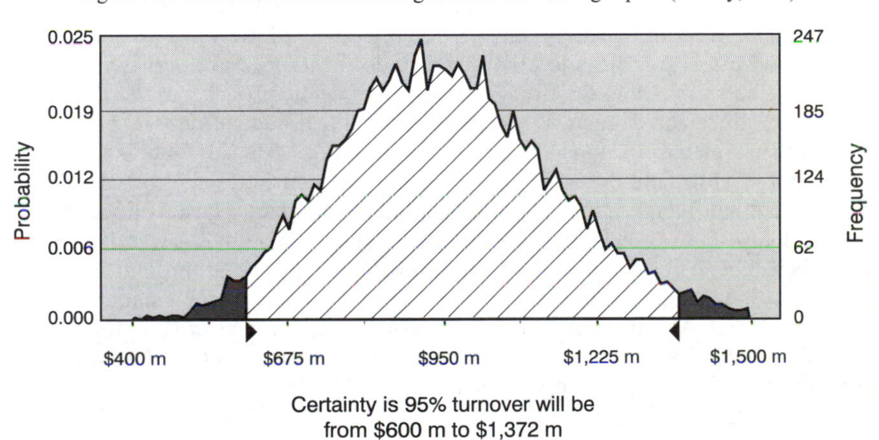

Figure 7.18 Forecast: Annual turnover (Kenley, 1999)

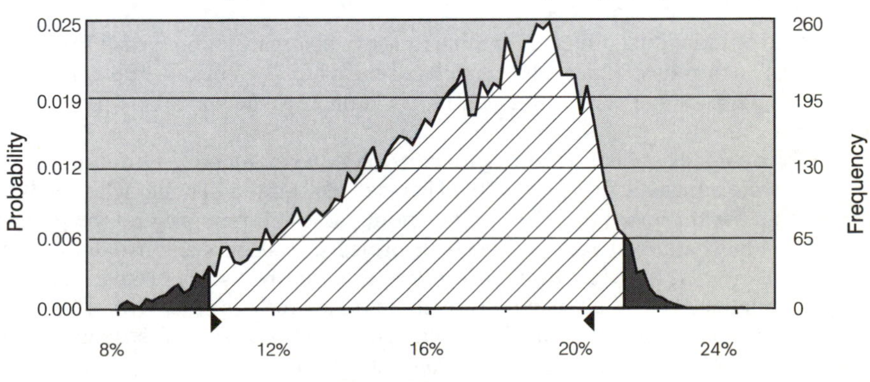

Figure 7.19 Forecast: Minimum working capital as a percentage of annual turnover (Kenley, 1999)

These results indicate that on average a minimum of 16.7% of annual turnover is available for reinvestment, with the range being between 9.08% and 20.71% at 95% confidence, and a minimum of 10.36% at 95% confidence.

Discussion

The results indicate that careful management of cash flow through operations is capable of yielding a significant contribution to working capital. With an average of 16.7% of annual turnover available, it is easy to see how an organisation could become over confident. This equates to $160 million for a company with a $1 billion turnover. This is the minimum level, so most of the time this would actually be exceeded. It would be tempting to see this money as profit, rather than trade credit, and to use it inappropriately.

More importantly, for a company that is deliberately and knowingly functioning in this way, the implications are significant. The findings show that the organisation has a 95% confidence of always having 10% of turnover available from operations. This is sufficient to allow investment in long-term alternatives such as property (albeit subjecting the firm to the vagaries of the property market).

There are clear dangers associated with such re-allocation—in particular it cannot be sustained following a change in circumstances. Thus, if there is a halt in the flow of projects to feed the cash flow, then the retained funds will be called on. If reinvestment has occurred then this may lead to failure due to lack of liquidity.

In contrast, an organisation that has the financial strength to sustain such a call on funds is in a position to make great use of the cash generated through operations. In this way, the building process may be seen as similar to the finance industry, whereby the purpose of operations becomes to manage the cash flow for reinvestment. This raises the question, whether some organisations build for the express purpose of raising funds through operations.

CENTRALISED MANAGEMENT OF CASH FLOW

A portfolio of individual projects combine to form the organisational cash flow. This means that each project cannot be considered individually, but must be considered from the organisational perspective. Yet this is counter to the way many projects are structured.

Project organisational structures for construction work tend to be dominated by the 'God project manager' (Figure 7.20). This is the all-powerful being who heads up a hierarchical team structure and through whom passes all reporting on the project. Such individuals are requested to drive their projects to success and usually they do this well. However, the view from head office may not be as impressive. Here the organisational needs may be seen holistically and they may conflict with the project manager's needs for the individual project. Under the hierarchical structure, head office cannot easily manage their needs from the project.

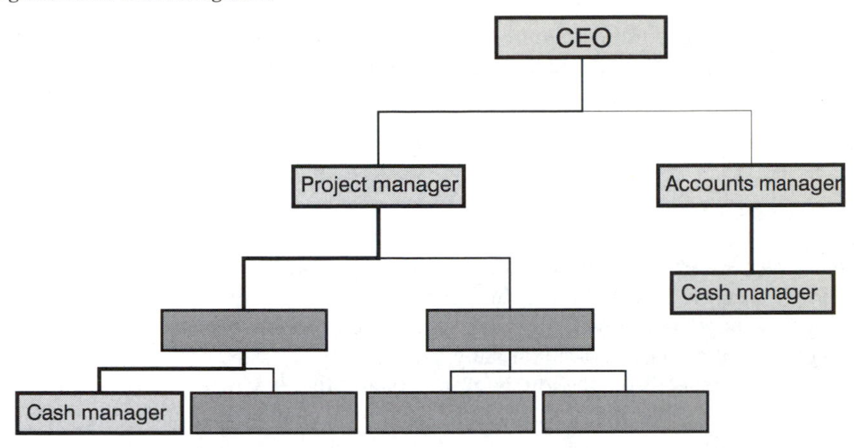

Figure 7.20 A hierarchical structure may subvert cash management needs

A matrix structure (Figure 7.21) may make it easier for head office to manage cash flow for the organisation. Such a structure is rare in construction because it leads to conflict with the project manager over his/her project, but it may be in the organisations best long-term interest.

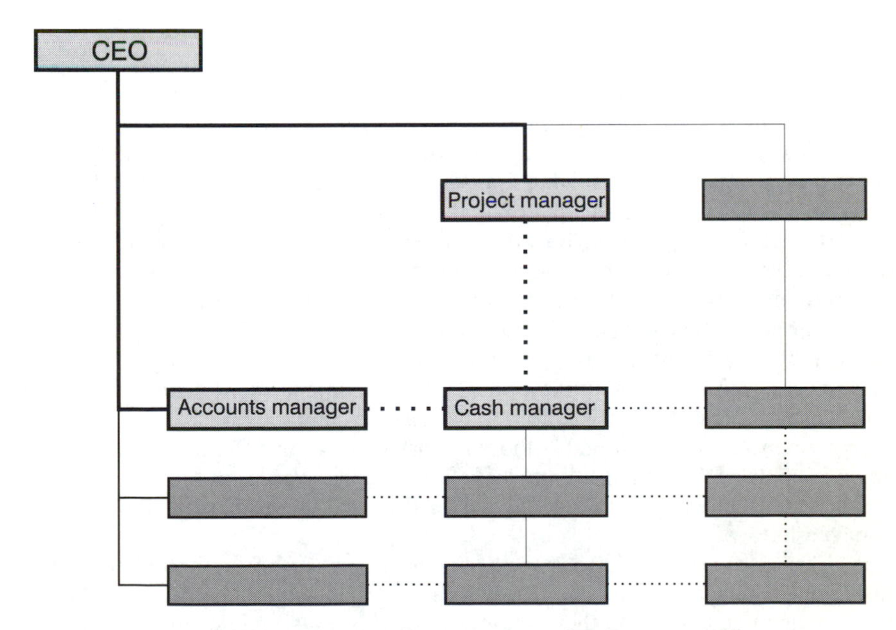

Figure 7.21 A matrix structure may support cash management needs

Whatever structure is maintained, it is important that the needs of the organisation in cash flow management are not subservient to the needs of the individual projects.

CONCLUSION

This chapter has explored organisational cash flow using a single set of assumptions to demonstrate the way that individual project net cash flows contribute to and form the organisational cash flow. This is one of the most important issues in the successful management of a construction company.

The key aspect to cash flow is to understand the way project cash flows layer over each other to generate organisational cash flow. A secondary aspect is to understand that organisational cash flow is dependent on the individual project profiles. Manipulation of these layered profiles has the capacity to change the organisational cash flow—which directly relates to the financial health of the organisation. This explains why cash flow at the project level is so important to success of the company.

Just as researchers have long indicated that focussing on cash flow is a key determinant in avoiding insolvency in construction, it can be seen that focussing on project net cash flow is the true factor for this success. In turn it is the overall management of a portfolio of net cash flow which determines the success of the whole organisation.

There are many alternative ways to operate a construction business, and that business may have objectives beyond construction. As yet the discussion of cash flow has ignored these more strategic aspects to project cash flow. It has been recognised that project cash flow falls within an organisation, that organisational strategy determines project net cash flow and that a portfolio of projects requires net cash flow to be a central function rather than a distributed function.

It remains to be explored how the strategy of the organisation can effect organisational cash flow. Such strategy could relate to structuring contracts to optimise organisational cash flow. Research could examine alternative strategies to measure their impact on funds raised through operations. Differences may be expected between managed contracts and guaranteed contracts for example. Alternative procurements systems such as turnkey, partnering and alliance contracting may significantly change the contractor's ability to use cash to fund their business.

Behind all this, and arguably more important, is the need to understand the relationship between the organisational cash flow and the success of the company. This requires understanding the path to failure for firms which mismanage their organisational cash flow.

Chapter 8 carries the discussion of organisational cash flow through to strategic cash flow management. Here the discussion of organisational cash flow is brought back to the interaction with the economy through supply and demand, the need for firms to generate cash flow through bidding strategies (such as competing on margins) is discussed and it is shown how these factors may lead to disaster. It is also shown that understanding such issues, and applying strategic management of cash flow, can contribute significantly to a organisation's success.

REFERENCES

Archer, S. H. (1966). 'A model for the determination of firm cash balances'. *The Journal of Financial and Quantitative Analysis*: 1–14.

Ashley, D. B. and Teicholz, P. M. (1977). 'Pre-estimate cash flow analysis', *Journal of the Construction Division, American Society of Civil Engineers*, Proc. Paper 13213, **103** (CO3): 369–379.

Bromilow, F. J. and Henderson, J. A. (1977). *Procedures for Reckoning the Performance of Building Contracts*, 2nd edition. Highett, Australia, Commonwealth Scientific and Industrial Research Organisation, Division of Building Research.

McCaffer, R. and Harris, F. (1977). *Modern Construction Management.* Crosby Lockwood and Stables. Chapter 10: 'Cash flow forecasting': pp. 166–183.

Kaka, A. P. and Price, A. D. F. (1991a). 'Net cashflow models: Are they reliable'? *Construction Management and Economics* **9**: 291–308.

Kaka, A. P. and Price, A. D. F. (1991b), 'Relationship between value and duration of construction projects', *Construction Management and Economics*, **9**, 383–400.

Kenley, R. (1987). Construction project cash flow modelling. PhD Thesis, Melbourne, Department of Architecture and Building, University of Melbourne.

Kenley, R. (1999) 'Cash farming in building and construction: a stochastic analysis'. *Construction Management and Economics* **17**: 393–401.

Kenley, R. and Wilson, O. D. (1986). 'A construction project cash flow model—an idiographic approach'. *Construction Management and Economics* **4**: 213–232.

Kenley, R. and Wilson, O. D. (1989). 'A construction project net cash flow model'. *Construction Management and Economics* **7**: 3–18.

Kenley, R. and Wood, B. M. (1996). 'A pilot study of the supplier/subcontractor payment system'. *Proceedings of Organisation and Management of the Construction Process*, CIB W65, Glasgow, **1**: 398–405.

McElhinney, K. (1988). Efficiency measures of working capital management for building projects. Masters Thesis (partial fulfillment). Melbourne, Department of Architecture and Building, University of Melbourne.

Nazem, S. M. (1968). 'Planning contractor's capital'. *Building Technology and Management* (October): 256–260.

Nicholas, J. (2000). A risk analysis system for evaluating construction contractors by potential creditors. PhD Thesis. Wolverhampton, University of Wolverhampton.

Paté-Cornel, M. E., Tagaras, G. and Eisenhardt, K. M. (1990). 'Dynamic optimisation of cash flow management decisions: a stochatic model'. *IEEE Transactions in Engineering Managament*, **37**(3): 203–212.

Punwani, A . (1997). 'A study of the growth-investment-financing nexus of the major UK construction groups'. *Construction Management and Economics* **15**: 349–361.

Warren, C. S. and Fess, P. E. (1989). *Principles of Financial and Managerial Accounting*, 2nd edition. Cincinnati, South Western Publishing Co.

8

Cash farming: Strategic management of organisational cash flow

INTRODUCTION

The discussion so far has been heavy going, with a lot of literature theory, and numbers. While endeavouring to maintain rigour, we now move into an area which has been much less the subject of research. There are only a few academic papers on strategic management of organisational cash flow, and these make assumptions about reality which are subject to dispute. There is little in the way of empirical work to either support or refute these assumptions and we are left floundering for strong arguments to provide understanding.

Underlying this work is the belief that the management of the timing of payments is central to the effective (or otherwise) management of a construction business. In the absence of empirical evidence, the role of cash management will be addressed using a reductionist approach. The case for the role of cash flow management in the life cycle of the construction firm will be constructed using logic and tested, where applicable, using simulation methods.

Cash farming is a light-hearted analogy deliberately used to describe very possibly the most important issue for those wishing to engage in the management of a construction business, the strategic management of organisational cash flow. The term 'cash farming' has been used deliberately to draw out the meaning implied in the term that we deal in this industry with the deliberate exploitation, to derive benefit, of the fallow field of construction cash flow. It is the milking of the cash cow. And, to stretch the analogy to breaking point, just as a real farmer can use and abuse the land, so can we see good and bad farm management practices at work in construction.

Cash farming is a term used by Gyles to describe a particular cash management strategy identified by the New South Wales Royal Commission into productivity in the building industry (Gyles, 1992). Such a practice was observed to have often been associated with company failure. The practice has therefore become blackened, as it is publicly associated with failure and not acknowledged by successful firms.

It is argued here that cash farming is neither good nor bad management. It is simply a management practice commonly used within the industry. Indeed, good cash farming techniques can contribute significantly to a firm's success and are too important to be ignored.

There is a light side to this topic. Imagine being in a room full of gruff senior industry contractor practitioners while canvassing these issues. Those that are still in business form one camp supporting the view (traditionally accepted in the research community) that payments lag work and income in a regular way and that no funds are deliberately generated through operations. Those that are retired and no longer protecting their firm's reputation, form the other camp, frankly admitting that the first camp were just plain dishonest in expressing their opinion. Such situations can become very intense and it is no wonder that researchers are wary of this issue as a research topic.

In this chapter, an attempt will be made to look afresh at this complex and emotive topic. Experience from presentations to industry demonstrates that strong reactions may be expected, with almost equal numbers in support and opposition. Perhaps the academic reviewer that once wrote 'this topic is not ethical and should never be published' will object most strongly, but support may come from the small builders who finally understand where their true profit lies, and with the silent army of subcontractors and suppliers who will quietly nod and say 'we have been saying this for years, why doesn't anyone ever listen'.

This issue explains why head contractors fail and how they make money, how they stay in business, and (at least one reason) why they use the subcontract system rather than direct labour. While the topic does not explain the entire industry, it certainly impacts on most aspects and is crucial for understanding underlying motives and mechanisms.

Most importantly, whether managers accept or reject the underlying values of this work, through understanding the complexity of interaction they will be better placed to form a competitive strategy for the financial management of their business.

BACKGROUND

The construction industry generally has low barriers to entry, and permits small, undercapitalised operators to enter and exit at will (Ashman, 1994). This encourages a culture where working capital is generated from operations, as firms struggle to overcome their lack of financial backing. There is a great deal of evidence that contractors manipulate the payment system in order to achieve this end. The Royal Commission of NSW identified the problem and illustrated it with case studies (Gyles 1992). Gyles analysed two significant failures: Girvan NSW Pty Ltd and the K B Hutcherson Group. In both cases, significant factors were found to be a lack of capital, and manipulation of funds made available through the progress payments system.

In the first case, it was reported that any cash flow generated by the construction division was utilised in lending to those parts of the organisation that required immediate cash flow. Such cash was used for both the construction and development parts of the business. Normal margins would not have been able to drive this re-allocation of resources; accordingly it may be derived that surplus cash was being generated through the timing of payments, and this was being reinvested. Such manipulation is common in the industry, but only sustainable while the positive cash flow is maintained and sufficient for the purpose. For Girvan, it was neither.

In the second case, the K B Hutcherson Group was financing itself by deliberately manipulating the cash flow of the projects so that the building owners, trade creditors and employees were unwittingly financing the operations of the group (described by Gyles as 'robbing Peter to pay Paul'). The liquidator noted that, largely, the company was funded by progress payments in excess of work in progress. The company's cash flow initially suffered from delays on projects caused by unusually wet weather! The cash flow shortage, together with underpricing on certain projects led, in many cases, to an inability to complete jobs. Further, the lack of new projects resulting from a decline in the construction industry denied the company any chance of trading its way out of difficulties.

Gyles adopted the phrase 'cash farming' to describe these practices and accepted that they are significant whether companies are solvent or not. It is interesting that research has so far largely ignored such practices, with the exception of attention to the practice of front-end-loading (Tong and Lu, 1992) which has seemingly been accepted as normal practice (Ashley and Teicholz, 1977; Kaka and Price, 1991; Kaka, 1996) despite the practice being ethically questionable and one which would most certainly be challenged by a client's quantity surveyor were they aware of its use. It is possible that this indicates a reticence on the part of industry to be identified with such practices (contractors generally claim to be good cash managers), and academics may not be eager to rock the boat as they are increasingly reliant on industry support or funding for their research.

Punwani (1997) identified from an analysis of company balance sheets that contractors generate working capital from their turnover. Punwani highlighted the importance of a portfolio of group activities, as:

- The reduced risk of simultaneously incurring downturns in all of the chosen sectors of activity; and
- The opportunity to divert surplus funds from cash-generating, low-investment-requiring businesses into cash-requiring, potentially profitable ones.

The suite of practices described by the term cash farming are endemic in the industry and company failure results from perpetrators being unable to accommodate changes in their financial circumstances. There are four factors at work.

1. An initial shortage in capital.
2. Funds for operations are derived from projects.
3. The funds available from operations are reallocated to either paying out losses or reinvested in non-liquid assets; and
4. There is a reduction in supply of cash flow, usually caused by a slowdown in supply of projects or delays in their execution.

If one of these points does not apply, then firms may, practically, continue to trade. For example, many companies either survive or sustain operations for a considerable period of time following a period of loss (bad projects) because they are able to maintain a steady supply of work. Indeed, failure comes about only when there is a downturn in the market (or one 'bad project'). This creates the phenomenon that may be observed whereby groups of companies fail during a sudden downturn. These firms may have been technically (and unwittingly) insolvent for some time and were hanging on, only to be tripped simultaneously by the same unfortunate circumstances.

The way that cash farming may contribute to failure of companies or, at least, determine the particular timing of the failure, is now evident. These are the temporary shortages to which Kaka and Price (1991) were referring. The issue is, however, more than a temporary shortage, but rather the deliberate reallocation of cash resources. In summary, cash farming:

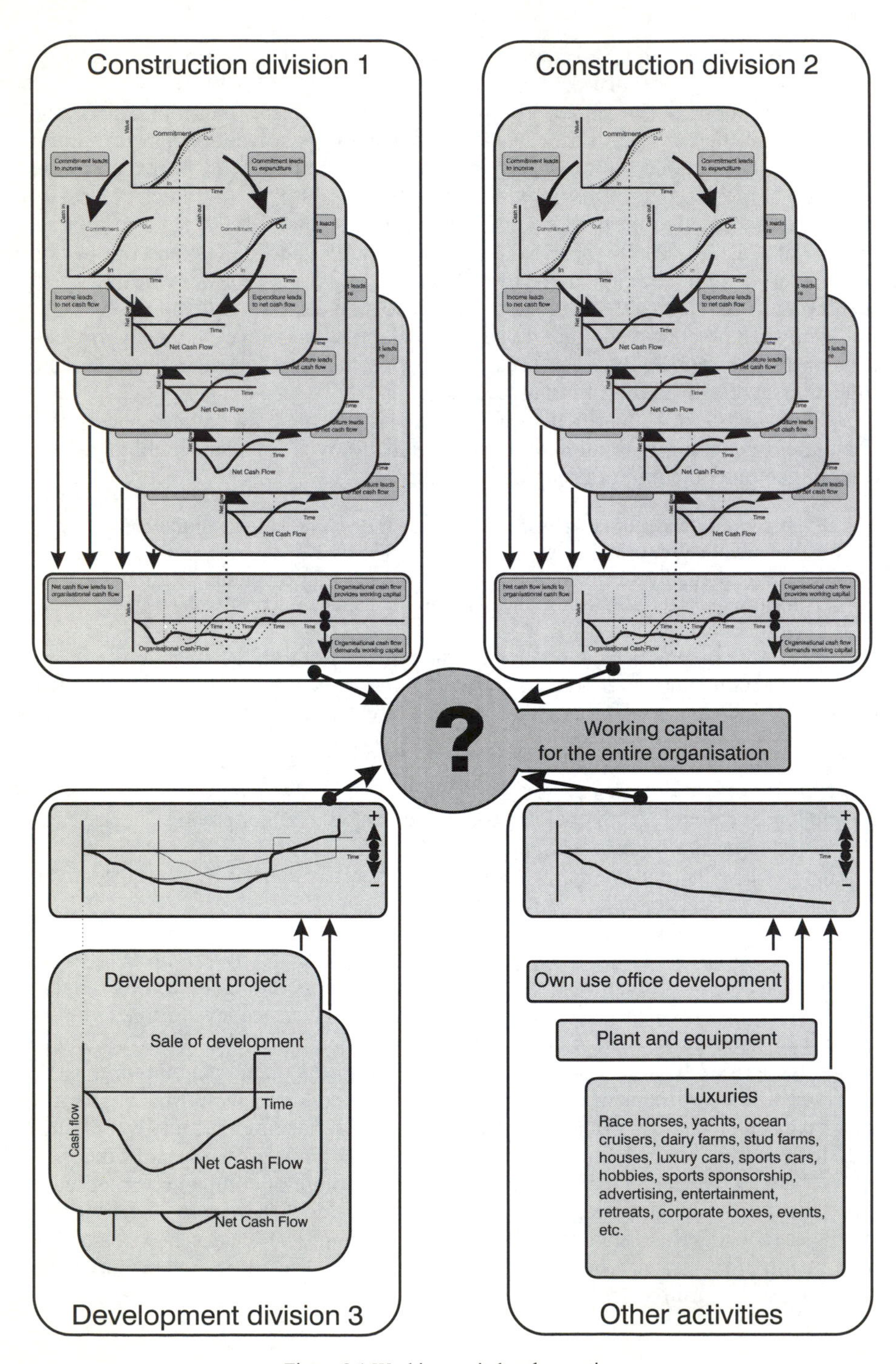

Figure 8.1 Working capital and operations

- Is a widely accepted practice in the building industry;
- Does not necessarily lead to failure on its own; and
- Can lead to a cash crisis, if investments are not able to be liquidated rapidly, following circumstances such as losses on a project, project delays or a failure to attract new work.

MANAGEMENT OF ORGANISATIONAL CASH FLOW

This chapter is about the strategic management of organisational cash flow, and through this, working capital required to fund operations. Figure 7.1 illustrated the way projects contribute to the organisational cash flow. However, this Figure says nothing about the conflicting business demands beyond the project level.

This more complex business context is illustrated in Figure 8.1. The total organisation as a whole requires funds for all its operations. While the construction component is likely to be significant, it certainly may not be everything. The business may have multiple construction divisions (for example, regionally based and separate operations). It may engage in development projects, each one requiring significant investment. It may be involved with businesses outside construction (but this is not really the concern here). It may wish to build its own head office, for partial use and further lease. It may purchase plant and equipment, possibly even establishing a plant hire business. Finally it may indulge in more frivolous activities of a marketing or luxury nature. This latter category can be truly fascinating, and can include: race horses, yachts, ocean cruisers, dairy farms, stud farms, houses, luxury cars, sports cars, antique cars, hobbies, sports sponsorship, advertising, entertainment, retreats, corporate boxes, events, etc.

Each of these components of the business may either deposit funds into, or withdraw funds from, the total organisational working capital. However, total organisational funds must come from somewhere. Where they are negative, funds must be raised by borrowing or equity. As noted, construction companies often have a very low capital investment, which leaves borrowing as the only source for many companies, plus the money that can be raised through operations.

The management of organisational cash flow is therefore critical to the operation of the business. If too many components of the business provide negative rather than positive cash flow, the net demand on resources may be too high to continue operations. Businesses must be strategic in their management, and establish a strategy which maintains positive, or at least manageable, net cash flow for the organisation. organisational cash flow

Further than this, it is easy to see that different components of the business have different timing for their demands and contributions to net cash flow. For example construction projects often require funds at the start and provide funds at the end. The amount required is generally not large because the client pays monthly, and most subcontractor payments are delayed until after receipt of payment.

Development projects usually place a greater demand on resources for a much longer period, not becoming a provider of funds until completion when units are sold, space is leased or the development is sold.

Some investments may never return a net positive contribution. These may be entered into for lifestyle reasons (horses, farms, boats, cars) or for marketing reasons (sponsorship, corporate boxes, events). Whatever the reason, they place a permanent drain on resources which must be offset with a corresponding contribution somewhere else in the organisational system.

Organisational working capital comes either from external sources, such as shareholder funds (equity), loans such as overdrafts (liability) or surplus cash generated through operations (assets). These seem simple enough, but there is another category, a hybrid, which may trap the unwary builder. This is the surplus funds generated, not through profit, but through the timing of payments allowing accrual of funds at bank but committed to sub-contractors or suppliers. This is actually credit being extended by suppliers and subcontractors, sometimes known as 'trade credit'. Through trade credit, one project which provides surplus funds may support another project which may be in the early stages and therefore demands funds.

It is possible for one division of a business which has a net surplus of funds to provide funds to another section which requires funds. In this way, a successful construction division which is generating funds may be able to fund developments, the purchase of further business or other organisational ventures.

The management of organisational working capital is the key to a successful and sustainable construction business.

Running multiple projects

A construction business is heavily dependent on the management of projects for its working capital, although that dependency might be both positive and negative. If we assume that one project places a small demand on resources as it starts and then eventually becomes a net contributor, it can be seen that a single project should rarely be a problem.

Difficulties arise when many projects are taken on. With them come the problems of a shortage of cash reserves needed to start more projects. Also comes the potential for apparently good times to be followed by a shortage of work, with an associated cash crises.

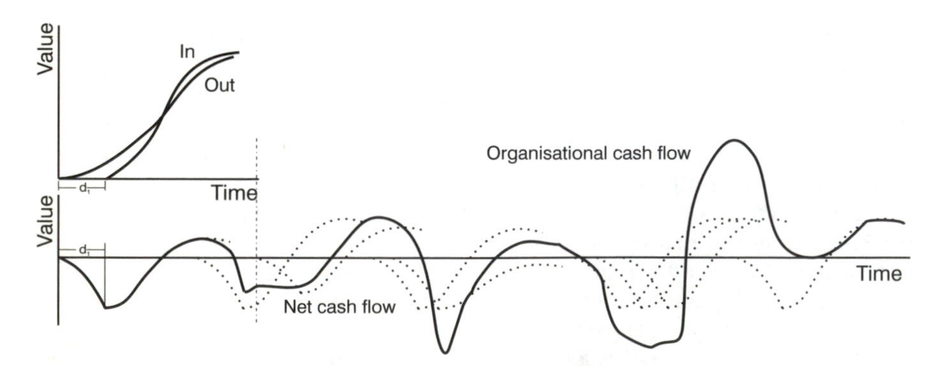

Figure 8.2 Diagrammatic representation of organisational cash flow assuming a traditional project profile, on a regular cycle (Concept first published by Nazem, 1968: Figure 2)

This is best illustrated using the traditional project cash flow profile and simulating the overlaying of many projects, similar to that developed in Figure 7.3 and illustrated diagrammatically in Figure 8.2. This cumulative profile assumes rapid dispersion of profit at the end of each project and so represents the contribution of funds generated (or absorbed) by projects. This indicates the need for working capital to be provided, or indicates the provision of surplus working capital. It can be seen from this figure that further sources of funds are required in the early stages of a project and that this need compounds when there are multiple projects. Such funding is normally drawn from overdrafts or similar facilities. It can be seen that starting new projects can be tough, and starting many such projects is tougher. Thus it is best that project starts are timed to coincide with project completions so that the one can fund the other.

Taking a more modern stance toward cash flow management, as per Figure 7.4, results in a very different organisational cash flow profile. This may be seen in Figure 8.3. Further sources of funds are rarely required, and organisational funds are predominantly positive. More projects yield greater funds. It can be seen that starting new projects is relatively easy, only presenting a problem when starting many projects simultaneously. Having many projects running is generally, however, very positive.

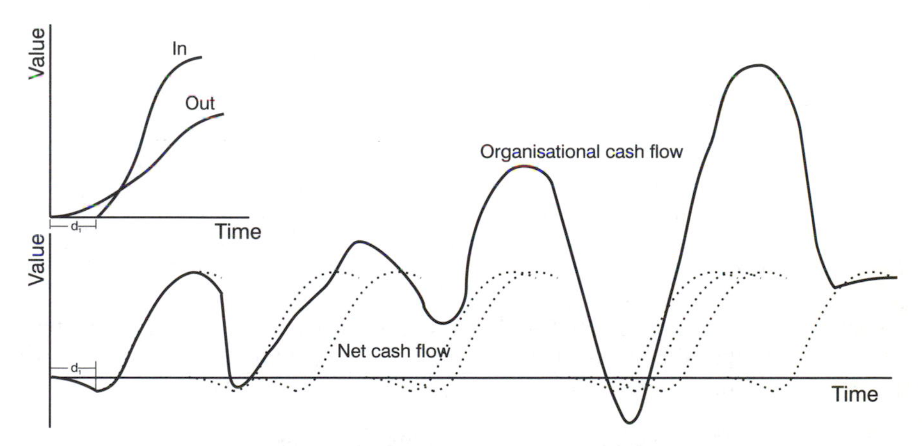

Figure 8.3 Diagrammatic representation of organisational cash flow assuming a modern project profile, on a regular cycle

A more aggressive stance toward cash flow management portrayed would result in positive cash flow throughout the life of the project. Similar to that portrayed in Figure 7.2, such an approach would yield a consistent and positive contribution to organisational cash flow. Starting projects is not a problem and funds are generated throughout the life of the project, compounding with multiple projects. Such a cash flow is illustrated in Figure 8.4.

These illustrations are diagrammatic and can be simulated using the mathematics of organisational cash flow outlined in Chapter 7. It is sufficient to understand that positive cash flow can be achieved early and sustainably on projects through legitimate means and within the contract terms. It is this overall positive cash flow, at the organisational level reaching as high as 30%, which is important to consider.

The difference between these representations is the role of management in managing each individual project. It is at the project level that a difference can be made. At the organisational level it is too late. However, central management can dictate the cash flow management policies to be followed for each project.

Before it is possible to examine strategy in cash flow management, it is necessary to recognise that companies have varying awareness of the true cash flow position. This is best illustrated by working through a scenario and demonstrating the consequence of management actions.

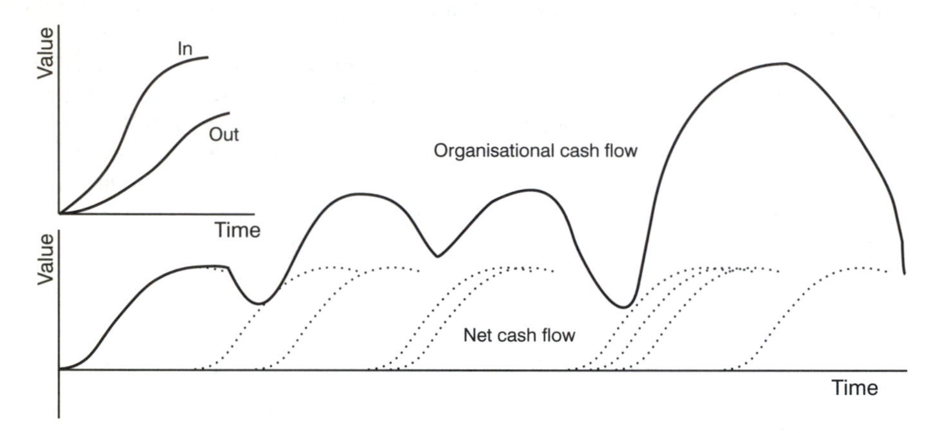

Figure 8.4 Diagrammatic representation of organisational cash flow assuming an aggressive profile, on a regular cycle

Types of firm awareness

A construction company operating within normal contractual terms, and paying bills on time (but to the strict contract terms), and not doing anything to manipulate the timing of payments, is able to achieve:

- **Minimum** funds in bank of 10% turnover, to as much as 20%.
- **Maximum** funds in bank of 20% turnover, to as much as 30%.

Although the funds generated fluctuate wildly during projects, and from month to month, contractors can become confident (complacent) about the presence of positive funds (Figure 8.5). They conclude they have become a very profitable business.

The process here takes two forms depending on the knowledge or needs of the contractor:

- **Knowledgeable** contractors who understand their cash flow position in detail would understand that none of the funds generated belonged to them. In this situation they would be aware that as projects wind down, they will be required to meet their commitments. They will make provision by either leaving the funds untouched, or more likely, ensure that any re-investment is easily liquidated or

replaced in the event of a change in circumstances. Let us call such an organisation the *'Aware Firm'*.
- **Less knowledgeable** contractors, or those whose needs or greed outweigh their common sense, are likely to not have detailed understanding of their cash flow position and have no ready explanation for the funds generated. In this situation they would be surprised at the state of their bank account, they would be delighted with their success, they might believe that, even if not all is profit (what a nice idea though), that at least a significant proportion is their return for all the hard work they have put in—after all they are a competent, well respected, firm. Let us call such an organisation the *'Surprised Firm'*.–

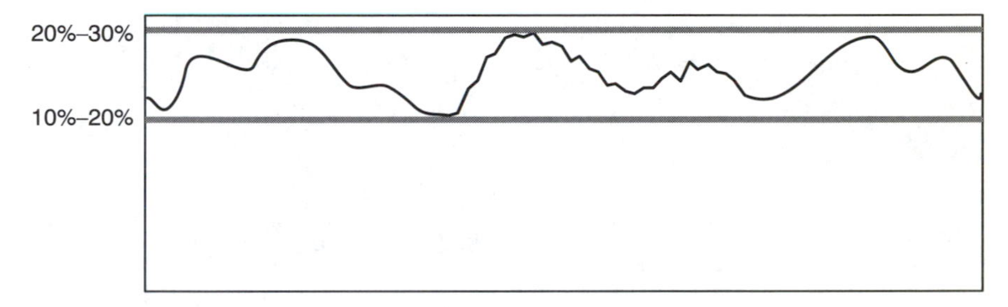

Figure 8.5 Organisational cumulative cash flow in happy circumstances—generating complacency

The *Aware Firm* is in a powerful competitive position, able to take advantage of their knowledge to deliver added value to their organisation. The *Surprised Firm* is in a weak position and highly likely to suffer business failure.

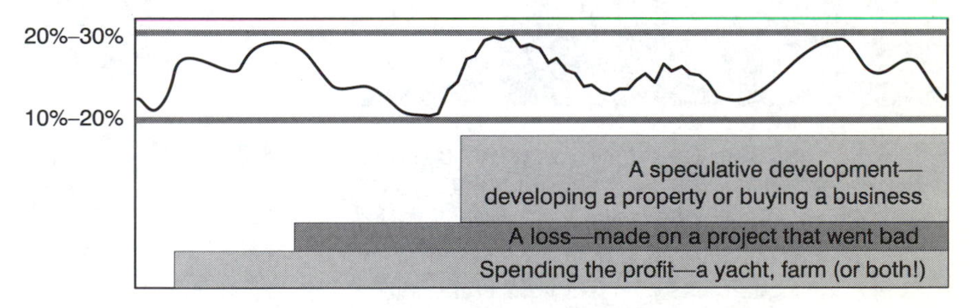

Figure 8.6 Organisational cumulative cash flow in happy but dangerous circumstances—spending with complacency

It is an axiom that when a firm fails, it was someone else's fault—usually the developer for late or non-payment. This is very unlikely to be the true cause, as failure to pay part of one project should never send an experienced and profitable company to the wall. In reality, their problem is likely to be a compound of other factors, and the missing payments become the proverbial straw to break the camel's back. Let us consider the path to failure for the *Surprised Firm* and then compare this with the path to success for the *Aware Firm*.

The path to disaster for the 'Surprised Firm'

The *Surprised Firm* does not fundamentally understand where their, sometimes considerable, money comes from. At this point we are in the territory of gross over-simplification, however, elements of the following illustrative tale apply in many cases. The *Surprised Firm* may consider the money as profit (which it is not). In this case they may, progressively, use that resource for other needs. Figure 8.6 illustrates the financial position of our hypothetical case.

Here, the owner, or the firm, initially purchases a yacht. In one recent example, the yacht was a 24 metre (80 ft) Forbes Cooper motor yacht called 'Risk & Reward'[1] shown in Figure 8.7.

Figure 8.7 The motorised yacht Risk & Reward, now available for rent

> Imagine…a luxurious white motor yacht on a sunlit, azure sea. Intimate anchorages with only the soft breeze sighing across the bay. Making ocean passages in

[1] Unfortunately, the author has no financial or other interest in the firms currently renting the yacht *Risk & Reward*. However, these images are provided for those interested in renting the yacht, and may be found at these web sites :
http://cruzan.com/yacht_charters/luxury_yachts/rreward.htm or
http://www.boatus.com/charterdir/3940.asp

> total comfort and safety. Your own captain, chef and crew. Exquisite catering. Fine wines. A complete selection of water toys. Experience the ambience that only good friends can bring. Going where you want to go, when you want. Freedom of choice. An adventure. **Risk & Reward**.

This is the dream, in this particular case now being rented in Ft Lauderdale, Florida, but once the proud dream of a contractor from Auckland, New Zealand. Such dreams may appear to serve a valid purpose. For example, the above yacht may have been extremely rewarding, even financially, during America's Cup yacht races. However such dreams tend not to be easily liquidated, particularly in difficult times.

Dreams come in many forms. For example, contractors will often own farms—as an investment. This is sometimes related to another dream, owning racehorses. A cynic might even argue that without the construction industry, the Horse Racing industry would not be viable. These are all investments, a way of investing profits into commercial opportunities, but ones which also realise a lifestyle benefit. After all, contracting is a risky business and to take the risk, one is entitled to a reward. There are, of course, many contractors that successfully manage to support their dreams through their business in the long-term.

It's a rosy picture so far, so what is wrong with it? Things go wrong, things which cause problems, leading to difficulty for our *Surprised Firm*. There are two main categories:

- **Financial impacts**
  Financial impacts are those which act to reduce the level of available funds, such as losses or reduced margins on projects, unexpected organisational expenses, delays in payments from clients or disputes leading to legal and related costs.
- **Timing impacts**
  Timing impacts are those which relate to the progress of the works, such as delays in progress or completing projects, or delays (failure) in winning new projects.

Each of these causes results in problems due to the inability of the *Surprised Firm* to liquidate the 'investments'.

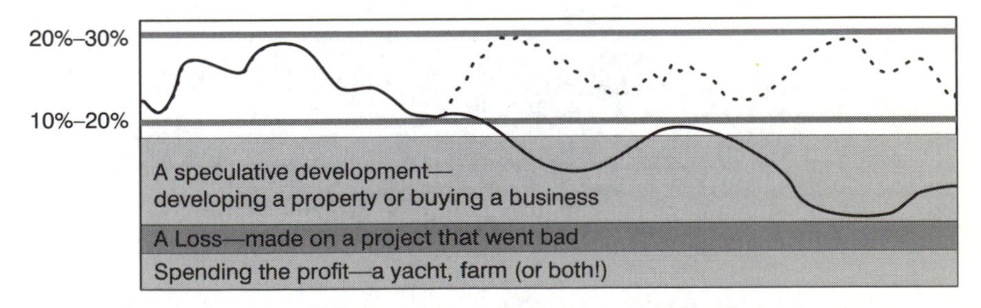

Figure 8.8 Organisational cumulative cash flow after spending with complacency: The impact of unfortunate circumstances

Whatever the impact, the effect is likely to be slow but inexorable. A downward trend in funds generated from projects will result as expenses are paid out, or as payments committed for work complete must be paid for without corresponding income for new work committed. This is illustrated in Figure 8.8, where the shaded area represents committed (and immediately unavailable) funds.

This circumstance is identified by Punwani, who noted from her analysis of company balance sheets, that:

> under conditions of recession, there is evidence of a change in direction of the working capital profiles. As total workload falls, there is a net reduction in the stream of interim payments from clients. Although the levels of debtors, stock and contract work in progress begin to fall, outstanding liabilities to creditors on old projects are still due. The creditor-financed working capital 'mechanism' described begins to break down; outstanding liabilities can be delayed for as long as possible but, ultimately, have to be paid (Punwani, 1997).

Clearly in these circumstances, the contractor needs to make up the shortfall. This can be done in the short-term by borrowing, usually by an overdraft facility. In the longer term, more drastic measures are required. At this point, given the difficulty in liquidating the 'investments', it is likely that management would decide that the problem is in fact a short-term effect of a downturn in the market or alternatively an intractable client.

The solution therefore would be to take short-term remedial action. Actions which a contractor might take include:

- Buy work—by reducing margins.
- Delay payments (temporarily) from 30 days to 60 or even 90.
- Dispute and make claims, take legal action.

These actions will have the effect of reversing the trend, by bringing in work and thus cash flow, by deferring payments and thus retaining cash flow, and forcing receipts and delaying payments through dispute. Figure 8.9 illustrates the resulting improvement in the organisation's circumstances.

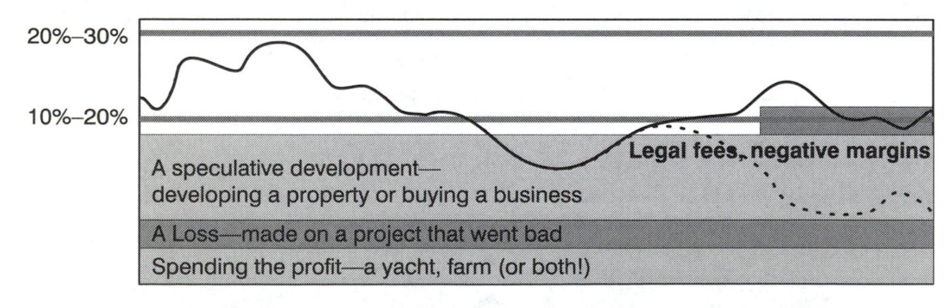

Figure 8.9 Organisational cumulative cash flow after acting to improve cash flow

Unfortunately, unless the cause of the downturn is resolved, this improvement is likely to be temporary. Buying work on reduced margins will lead to a loss of profitability, delaying payments will lead to a loss of support from suppliers and subcontractors, and disputes with clients will lead to a loss of competitive advantage when pursuing new work. Despite this, there are examples of contractors following these strategies for years, possibly even trading out of their problems through windfall gains or tight management.

A contractor experiencing the conditions illustrated above would be able to continue trading but would be finding conditions tough. If they can hold on until completion of the speculative development (assuming it is a property development rather than the purchase of a business) then conditions would ease. Such a contractor would see the sale of the investment as the solution. They might be right, and a profit on the sale would ease matters considerably. A loss on the sale, quite likely in the event of a financial down-turn, would not be quite as effective.

The improved apparent financial position, with restored positive cash flow and surplus funds, is illustrated in Figure 8.10 where the speculative development has been sold at a loss, the yacht is retained and costs of business in the form of reduced or negative margins and legal and dispute costs are gradually increasing (these are cumulative and can never reduce once spent, but rather must be paid for, through future profits).

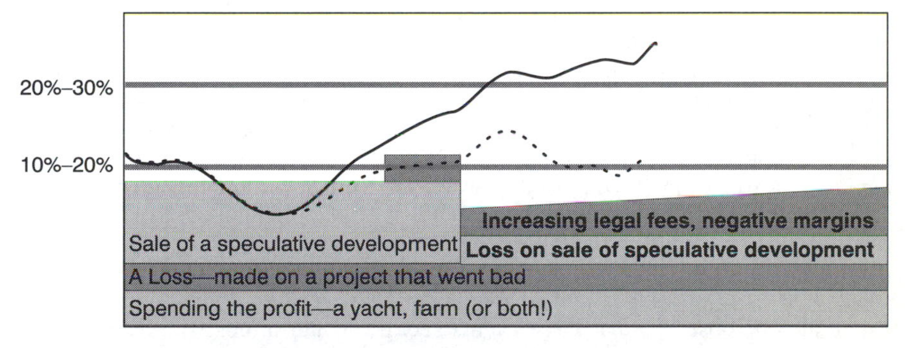

Figure 8.10 Organisational cumulative cash flow after acting to improve cash flow and selling the speculative development at a loss. Operating costs are increasing.

It is likely, however, that this position is not sustainable. In part the surplus is generated through delaying payments, but these eventually fall due. Incoming projects are becoming harder to win, due to increasing project costs and inflated supplier prices. To make matters worse, reduced quality as part of a cost cutting drive has made clients unwilling to pay and subjected the contractor to claims for defects (Cooke and Jepson, 1979). These problem projects then draw resources away from other projects, causing the effects to mount.

Eventually all these problems compound, usually quite quickly. With insufficient projects entering the system to generate new cash flow, and with committed expenditure eventually flowing out, the end is sudden, seemingly 'unexpected', and very costly to those to whom money is owed. Figure 8.11 illustrates the demise of our *Surprised Firm*. Clearly the responsibility for this disaster does not rest with the client who was reluctant to pay those claims, however, the blame will be laid at the feet of all clients and they may be vilified, as their reluctance to pay will be portrayed as being responsible for destroying an otherwise good business. It will not just be the contractor who will fail, the effects will also damage or destroy many suppliers and subcontractors.

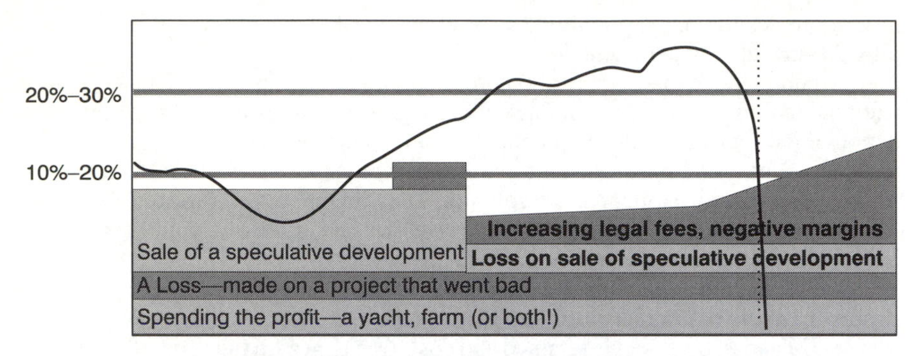

Figure 8.11 Organisational cumulative cash flow showing eventual and rapid failure

It is necessary to look at the impact of this disaster. If we assume our *Surprised Firm* is a small- to medium-sized operation with a turnover of say $200 million, then failure is likely to leave between 10% and 25% of turnover owing to creditors. For our company, this would leave owing between $20 million and $50 million. Such a failure would have a significant impact on the downstream companies in the *Surprised Firm*'s supply chain, as most of this money would be owing to them in what Davis (1991) refers to as the 'knock-on effect'. The knock-on effect works both up and down the supply chain, causing the insolvency of the dependent subcontractors but potentially also the client, who may be prevented from achieving their goals through the failure of the contractor.

Simultaneous failures

It may also be observed that these events seem to occur almost simultaneously in several companies. This is an interesting situation observed most starkly in 1990 in Victoria, Australia, during a period of severe contraction following the excesses of the late 80s. Private sector work was extremely rare at the time (at one stage it was rumoured that there was only one project available for tender in the entire city of 2.5 million people) and government work was supporting the sector. Joan Kirner became Premier in September and in an effort to stem the tide, soon froze government expenditure on capital works—including continuing contracts. The effect was rapid and disastrous. Over one notable two-week period, on average at least one medium to large contractor failed every day! This does not necessarily imply that these companies were all like our *Surprised Firm*, but it does show that they were living on trade credit (their suppliers and sub-contractors funds) and were reliant on continued new work for survival. Once the new work dried up, or the existing work was stopped, committed payments continued and companies failed.

Included in these failures were several significant organisations, including one which seemingly attempted to stem the tide through an aggressive advertising campaign (amusingly the theme was the image of a bomb with its fuse burning—probably to give the impression of meeting deadlines, but subsequently taking on a new meaning as the company imploded). Apparently this company also attempted a variation on the *Surprised Firm* strategy, they had tried to create work for themselves in the earlier recession through funding developments which they then built. This high-risk strategy indicates that this firm probably fitted the category of an *Aware Firm*, but one which failed to cover the need to be able to liquidate investments. They discovered just how little a partly finished project is worth on the market, they also discovered that creating work where you provide your own finance does not assist the cash flow needs of the organisation in the way winning other work (even at negative margins) would do.

Another organisation which failed at this time had been on an expansion drive. They were a company based in New South Wales but wishing to compete nationally, and had purchased a Victorian company to be the new regional division, using funds generated on construction projects to fund these purchases—and in turn needed further projects to support the expansion. With a general market downturn and the failure to win new work, the financial structures failed. Once again this was probably a deliberate strategy, but one without sufficient safeguards.

If failure is a likely result of progress for a *Surprised Firm*, what is the likely outcome for a well managed *Aware Firm*? The actions are likely to be very similar to the *Surprised Firm*, but the outcome is potentially very different.

The path to success for the 'Aware Firm'

The *Aware Firm* understands the generation of funds through operations and their responsibility to those to whom payment is due. This does not, however, preclude utilising those funds for investment. In this behaviour they are similar to a bank or other financial institution.

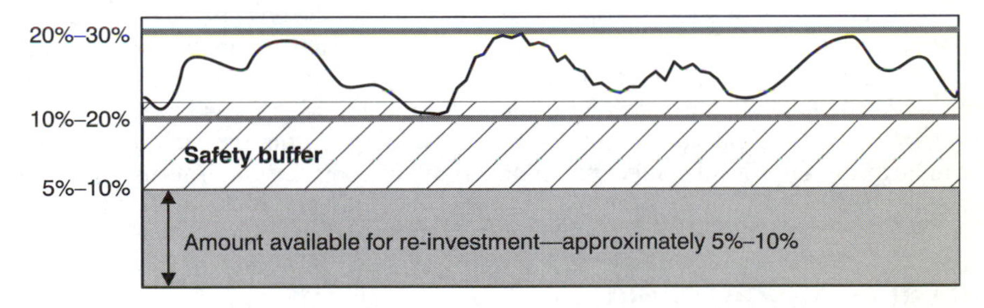

Figure 8.12 Safe reinvestment of organisational cumulative cash flow

It might be argued that it would be irresponsible, a clear example of waste, for a firm to hold 20% of its turnover without doing anything with it other than earning interest. The waste represents the opportunity cost of the full potential earning capacity. Figure 8.12 illustrates the likely range available for safe re-investment. Figure 8.13 shows that situation during a market downturn. Despite a short period when overdraft or similar funding is required, the firm is well placed to survive until the situation improves.

If we assume that our *Aware Firm* has a turnover of $200 million, then it could expect a minimum amount of funds generated to be between $20 million and $40 million. Furthermore, a reasonable assessment of risks would yield the likely call on funds in the event of a downturn. The resulting amount is available for re-investment. If we assume that this firm has a minimum of $30 million and that half that is required for a downturn to give enough time to release emergency funds, then $15 million is available to invest. This is sufficient to undertake a project equivalent to 7.5% of their turnover. This provides a source of work, the potential for significant profit, and lastly a mechanism to smooth workloads—allowing increased efficiency within the organisation. Assuming a conservative developer's return of 15% of project value, the firm might gain a profit of $2.25 million. It would take them more than four projects to earn the same profit given a constructor's margin of say 3.50%. This is powerful leverage for a construction business which is likely undercapitalised. Such action increases return on capital significantly.

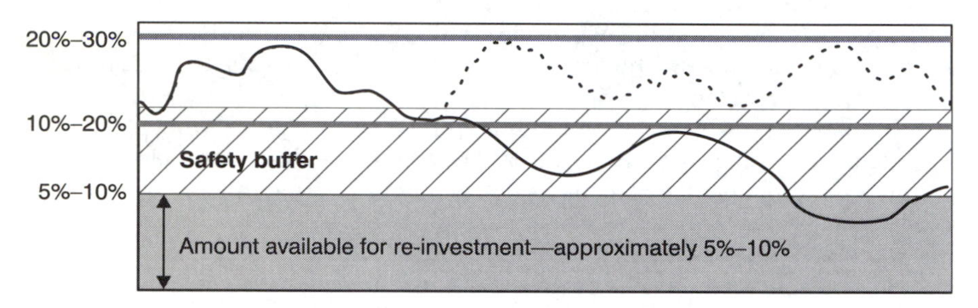

Figure 8.13 Safe reinvestment of organisational cumulative cash flow, despite difficult times

There is a risk here that a downturn will place a demand on funds, at a time when the value of the investment for which the funds have been used will also reduce. This should only be a minor inconvenience, in line with the normal risks of business investment, as long as the firm is not forced to retrieve value prematurely. However, the carcases of projects that stopped through lack of funds can be found in many cities. These failed before reaching a point where their value could be realised. It is generally very difficult to sell an incomplete building. Failure of such investments can be a very significant component in the failure of construction firms that invest in development.

CASH FARMING TECHNIQUES

Implications of the organisational cash flow simulation

It can be seen that a firm which is in a sound financial position may be able to improve its financial performance through careful cash management, or cash farming. If this were not the case then such practices would be rarely used.

A building or construction firm is able, through operations, to generate significant free positive cash flow, which may be used as working capital for purposes other than operations. This re-allocation of funds is sustainable, providing it is deliberate, carefully managed and conducted in a cautious way.

In Chapter 7 a simulation was used to explore the nature of organisational cash management, and the levels of working capital generated through operations. The stochastic method simulated real world circumstances, replicating management practices. The stochastic approach allowed investigation of a base range of organisational policy assumptions and calculation of the working capital so generated.

One reason to prepare this simulation was to test the belief that large building firms can generate sufficient working capital through operations to allow re-investment in further capital works. Anecdotal evidence indicated that a firm turning over $1 billion can generate working capital in the region of $200 million through their operations, funds which will not be called upon by those operations and which may therefore be safely re-invested. This equates to 20% of turnover being available for re-investment.

A model for the cumulative working capital of an organisation was presented under a range of financial policy assumptions. Implicit in these was the assumption that these policy decisions were divorced from functional issues, such as sub-contractor goodwill, dispute resolution, etc. The model also ignores any impact that such practices may have on sub-contractors and suppliers, although this would be minimal in the study, as it was based on accepted standard practices without applying any additional cash farming techniques.

The results of the simulation indicated that careful (but not extreme) management of cash flow through operations is capable of yielding a significant contribution to working capital. With an average minimum of 16.7% of annual turnover available, it is easy to see how an organisation could become over confident. This equates to $160 million for a company with a $1 billion turnover. This is the minimum level, so most of the time this would actually be exceeded. It would be tempting to see this money as profit, rather than trade credit, and to use it inappropriately.

Strategies

The analysis to this point has not included any unethical practices. This does not reflect reality. It has already been noted that the research community has accepted the practice of front-end loading. However there are other practices which affect the levels of funds generated from operations. These are discussed briefly here.

Front–end loading.

> The unbalancing of a bid occurs when a contractor raises the prices on certain bid items and reduces the prices on others so that the bid for the total job remains unaffected (Ashley and Teicholz, 1977).

Unbalancing is a highly dangerous and questionable practice. It is dangerous for the contractor's internal cost control mechanisms and it is dangerous for the client. Yet it seems to be accepted practice in the industry.

The danger to industry comes through errors in letting contracts. To illustrate, the following story comes from the experience of being commissioned to locate the cause of a series of project losses in a very old, very traditional building company, a company which subsequently failed (of course, blaming a developer for failing to pay, but that is another story). The context was as follows:

There were three projects of approximately $6 million. These had each suffered a loss at the conclusion of approximately $1 million. This was a serious problem for the firm. An investigation revealed the following situation:

- Rates were loaded onto early trades such as excavation, structure and roofing to get income early in the project in excess of expenditure.
- The early trades were loaded by about 20%, but the balancing was unequal, with a greater negative loading of about 30% for a shorter period in the late stages of the project.
- This information was restricted within the organisation and not revealed to the project administrator.

At about 60% of the way through the project, it became clear that there were problems finishing the project, but it was only at the end that the true extent of the damage became known. The investigation discovered:

- Early trade contracts that should have been let at a margin of 25% were let at 5%, which was the target margin for the whole project.
- This letting practice committed a significant proportion of the project at 5% margin.
- The administrators were unaware of the problem, as project reports declared the project on target.
- Once the error was discovered, at about 60% through the project, approximately $600,000 to $800,000 had been lost in letting errors.

To make matters worse, the reduced rates to the client for the finishing trades were so cheap, that the client requested additional finishing work. Additional landscaping, additional carpet in lieu of vinyl. All these additional works could only be let at a significant loss. These later letting losses compounded on the earlier to total close to $1 million.

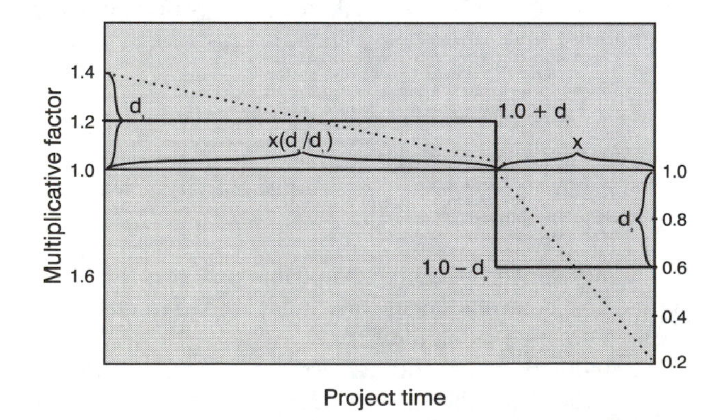

Figure 8.14 Non symmetrical unbalanced loading of margin for front-end loading

Ashley and Teicholz (1977), Kaka and Price (1991) and Kaka (1996) do not question this practice. In fact they accept a loading ratio of d_1=40% (refer Figure 6.20)

but they allow for a continual reduction through the project as if each cost item was a separate letting item and able to be variably adjusted (Figure 8.14).

In reality this is not realistic. While computer software could theoretically match the Bill of Quantities items to the work schedule and therefore allocate a sliding scale of loading, this is probably never done. Usually the variable loading is applied in blocks of items as trade or letting packages. Were these packages small enough, then the Ashley and Teicholz approximation might be reasonable. Rather than becoming easier on the sub-contract and supply packages let side of the equation, it actually gets harder, as the above tale highlights. Contracts tend to be let in blocks of time for the best available price. There is no necessary relationship to the letting margin, but errors in managing the allocation of loading become sensitive.

Figure 8.15 illustrates the mark-up practices applied to the client priced items in the above case.

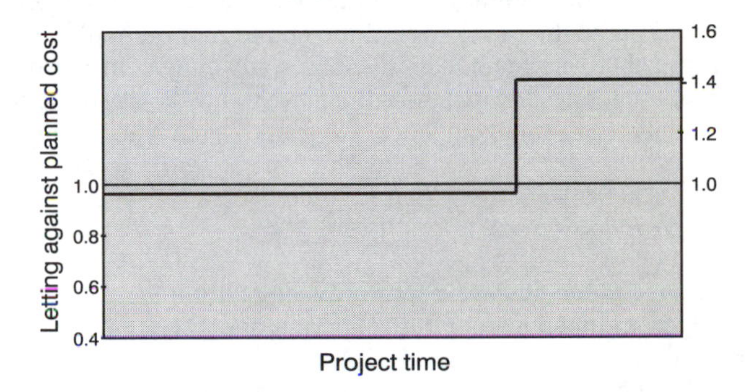

Figure 8.15 Consequences of front-end loading with poor letting—perceived margin is lost in the latter stages of the project

These graphs reverse for the letting side, relative to planned. If the project was run well, all packages would be let at the zero margin rate, and the loading would take place on the client rates only. Correctly let, therefore, a project would display a project margin against target of (say) 5% less than target throughout the project. In this situation, the project would display the correct margin for the first 60% of the project, and all would seem all right, but then would display irretrievable cost-over-runs for the balance of the project (see Figure 8.15), rapidly resulting in a significant loss for the project.

The advantage of front-end loading rates clearly outweighs the risks for many contractors. The advantage is one of cash farming. By loading the early trades, the contractor receives from the client a disproportionate payment relative to the value of the work in progress. The receipt payment curve is, effectively, pushed forward. This results in earlier payment from the client without affecting the timing of the payment out to suppliers and sub-contractors. This has the capacity to make an enormous impact on the net cash position, far greater than many of the issues thus far discussed.

The client also has a risk associated with this practice. By paying too much for early work, they are at a far greater risk in the event of failure of the contractor. It is never easy to complete work after a builder has failed, but the task is made even harder if it is discovered that the value of work to complete is less than the amount paid, thereby leaving insufficient to complete the work.

Delaying payment until requested

Delaying payment using the bottom drawer is usually spoken about with a wry smile, as most people acknowledge that the practice is not ethical. The practice is very simple and, believe it or not, quite common. The method is to place all the payment cheques (in their envelopes) into the bottom drawer instead of the mail. There they wait until the first (or second—or third) call requesting payment, at which time they are mailed. If done on the first call, the result is that generally no one gets upset. Contractors are of course not alone in this practice, many other organisations such as consultancies follow this practice. There is an easy test for it: those practising the method will never pay until reminded, whether you wait a week, a month or longer, although some will purge the file drawer at the end of the following month.

Apparently, using this method has another advantage. In the event that the supplier or subcontractor never calls, the bill is never paid. This is more frequent than would appear likely, anecdotal evidence suggests it may be as high as 4% of payments.

Engineering the cash flow stream

In reality, the above strategies are potentially more trouble than they are worth, being either dangerous, unethical, or both. There is another, simple, way to generate positive cash flow.

So far we have concentrated on traditional forms of contract by default. There are, however, firms who do not generally tender for work and who prefer not to enter into traditional contracts. A firm which is strong enough in reputation and which is able to direct the terms is able to re-write the strategy. In fact, it may well be that cash farming is sufficient reason to recreate the contractual and business relationship

There is a suite of contractual arrangements that, depending how they are constructed, allows the contractor to engineer the cash flow stream. Such arrangements as *guaranteed maximum price* and *design and construct* inherently contain mechanisms to engineer the cash flow stream.

The method, once again, is simple, although implementation may be a different matter. All that is required is to agree up-front with the client a guaranteed claims schedule to go with a guaranteed maximum price (or similar arrangement). There really is no sound reason for client payments to be tied to the cost profile in any direct way, apart from securing value for the payment through work in place.

Separating the value curve from the income curve is the key to sound project cash flow management, although it does require a strong firm with a trustworthy performance reputation and sufficient resources to guarantee performance. All the firm must then do is ensure that they do not build faster than the forecast cash flow which would match the claims profile. A diagrammatic representation of the separation of the income and value curves illustrates the power of this method (Figure 8.16). To ensure that this in turn does not concern the client, there should be a schedule which matches the real rate of progress.

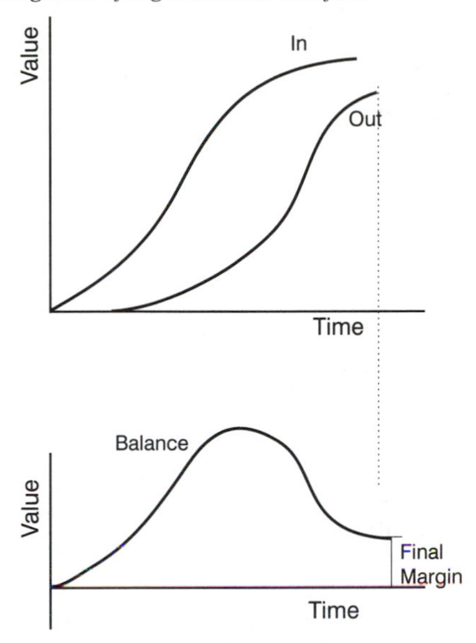

Figure 8.16 The separation of the income and value curves

With this technique it should be possible to ensure that:

- The initial dip into negative net cash flow is avoided at the start of the project.
- The project is a significant contributor to organisational cash flow throughout its life.

The irony here is that these alternative contract forms, while carrying a significant design and cost risk, come at a higher premium. That which makes sense for cash flow management, effectively also pays better. Margins move from 3% to 6% to 15% to 25% before considering the interest earning potential of the cash flow stream.

A side benefit of this method is that, due to the separation of the relationship between the payments from the client and the payments to sub-contractors and suppliers, the contractor is free to favour those sub-firms with whom they have established a long-term relationship. The pressure to delay payment in order to ensure their own financial viability is relieved. This provides a good framework for better relationships which would benefit all parties in the long-term.

It is probably fair to say that large successful firms which include development or financial divisions have long ago mastered the art of engineering the cash flow stream.

Up-front payments

Another way to achieve better relations among parties, and to ensure more chance of timely payment to subcontractors, is for the contractor to be paid in advance. If the client were to pay for work progressively but in advance, then the contractor would be able to pay sooner without running into cash flow problems. Without early payment to sub-contractors and suppliers, progressive payments in advance would significantly increase the net cash flow. This would give the same effect as shifting the income curve one month in advance, thus increasing the positive vertical difference between the two cash flow curves.

Khosrowshahi's negotiation approach

Khosrowshahi (2000) advocates 'a radical approach to cash flow management'. This is in fact a proposal that the client and contractor negotiate 'an inter-relating trade-off' between the parties, for example trading off better payment terms for a reduced project margin. In traditional contracts such a proposal has considerable merit, however much of the payment system is hidden and firms are likely to be reluctant to go too far in these negotiations, particularly contractors who are already in a very strong position.

Implications of the organisational cash flow simulation

In Chapter 7, a simulation of the overall cash flow for an organisation was developed. This showed that significant funds could be generated through adherence to the terms of existing contracts. It can be expected that using strategic cash management techniques, such as the dubious methods of front-end loading, deferring payments, extending payment terms, etc. would significantly increase this contribution.

Removing the connection between the inward and outward cash flow streams would provide a similar boost to total organisational cash flow. As this approach is completely ethical, this was tested in a revised model.

Revised model parameters

The revised model's assumptions were that the inward cash flow curve for a project would have a fixed profile, regardless of the progress of the works. The outward cash flow, in contrast, would be determined by the same stochastic process as used in Chapter 7. The α and β parameters were not adjusted for the outward cash flow from that used in Chapter 7. The inward cash flow used a generous profile, where $\alpha = 1.0$ and $\beta = 1.6$.

The variability of the time performance parameter K was reduced to a standard deviation of 30 (was 90) about a mean of 280 which indicates improved time performance and reduced variability. Similarly the Index ratio r was adjusted to January 2001 (BCI = 847) giving $r = 0.1677$.

To reflect the different business market that this organisation would be addressing, the assumptions regarding holding payments were altered, although retention assumptions remained the same. The holdings were accordingly reduced for months 1, 2 and 3 from 85%, 75%, and 30% (Table 7.6) to 85%, 50%, and 5% respectively (Table 8.1). A fixed monthly over-head of $2 million was placed on the organisation, off-set by 5% retained project margin.

Table 8.1 Adjusted holdings per month

Parameter	Chapter 7	Revised model
K, base 142 (1972)	$313 \pm 90\sigma$	$280 \pm 30\sigma$
B	$0.30 \pm 0.15\sigma$	$0.30 \pm 0.05\sigma$
Inward α	$0.22 \pm 0.74\sigma$	1.0
Inward β	$1.64 \pm 0.47\sigma$	1.6
Outward α	$0.22 \pm 0.74\sigma$	$0.22 \pm 0.74\sigma$
Outward β	$1.64 \pm 0.47\sigma$	$1.64 \pm 0.47\sigma$
Month 1	85%	85%
Month 2	75%	50%
Month 3	30%	5%

A typical project in the simulation is illustrated in Figure 8.17. This shows the separation between the inflow and the outflow curves and the resultant positive cash flow. While the stochastic analysis results in individual project variations in the outward cash flow profile, the inward remains a constant profile. The resultant net cash flow, however, does vary, and is sometimes negative (this is something which management would strongly discourage) (Figure 8.18).

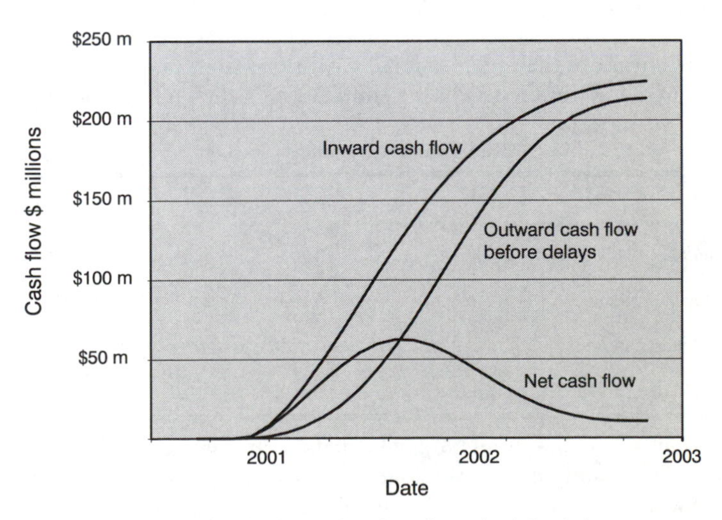

Figure 8.17 Indicative project from the revised simulation

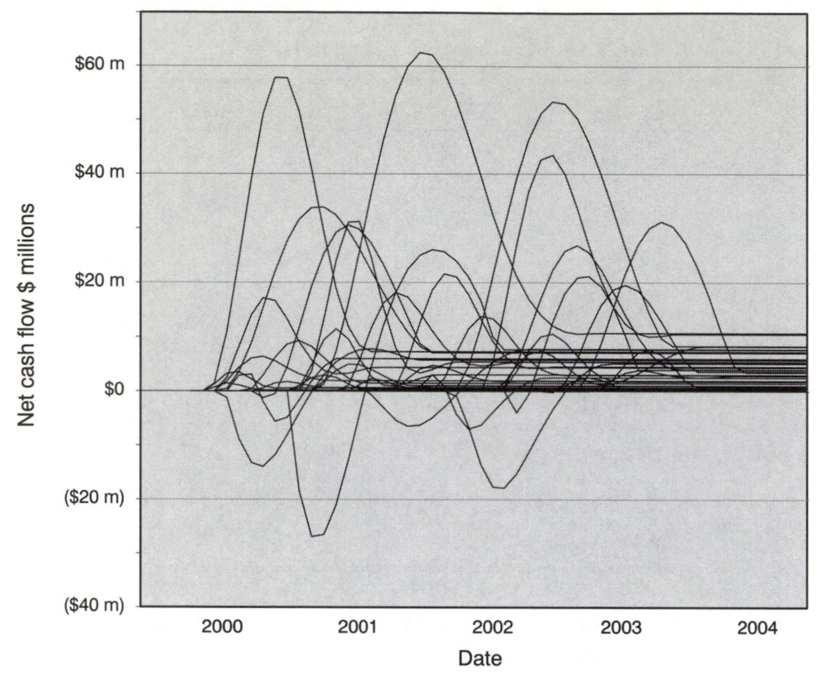

Figure 8.18 A set of net cash flow results from only one of the 1001 iterations in the revised simulation

Results of the revised analysis

An indicative result of the organisational cash flow, from one iteration of the 1001 which were explored for the stochastic simulation, is shown in Figure 8.19.

Table 8.2 Adjusted holdings per month

Indicator	Mean, σ	Range at 95%
Minimum funds available	$562 $\pm$ 171σ million	$267–$933 million
Average funds available	$703 $\pm$ 187σ million	$380–$1,127 million
Minimum funds, % turnover	55.6% $\pm$ 14.2%σ	28.3%–83.5%
Minimum interest earned as % of turnover	6.90% $\pm$ 1.45%σ	4.41%–10.0%

The results from the simulation indicate that the outcome is very favourable for a company. The minimum cash available for reinvestment is $300 million with 95% confidence which is minimum 32.5% annual turnover. A minimum 4.7% of turnover is earned as interest at 10%. Table 8.2 summarises the relevant organisational cash flow performance statistics. These results strongly support suggestion that improved organisational performance is to be achieved from separating the inflow and outflow profiles.

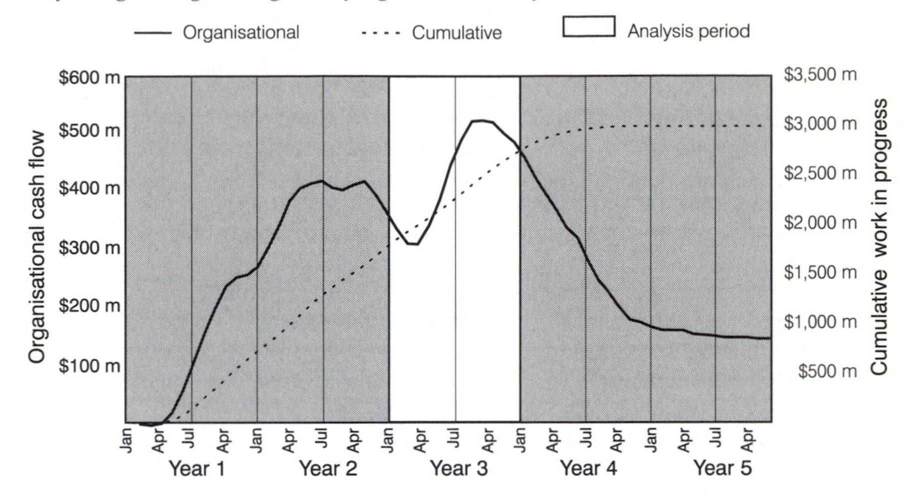

Figure 8.19 An organisational cash flow result from only one of the 1001 iterations in the revised simulation

This strategy does require a movement away from traditional thinking about projects, contracts and relationships. It is, however, ideally suited to Design and Construct, Guaranteed Maximum Price and Alliancing contracts.

DIRECT USE OF FUNDS

One aspect to cash flow management which directly relates to the timing of payments is the investment of funds received in short-term investments such as the money market. This is not really an issue of net cash flow but rather the direct interest earning potential of the cash flow stream, but does involve strategic management.

One of the more surprising 'games' played in the industry highlights this use well, one that can take place at any time, but particularly during periods of high interest rates such as the rates in excess of 18% in the mid 1980s. This game revolves around the timing of individual progress claims, which are often very large sums on major projects. The client wishes to defer expenditure for as long as possible as they have their funds invested at short-term rates. Equally, the contractor wishes payments received as early as possible. In this game, the client, in an attempt to delay payment for a few extra days, will deliver a cheque at 5 pm on a Friday afternoon, believing that funds cannot be drawn until Monday morning, giving them approximately two and one half extra days interest on the funds. However, in this game, the contractor reaches a special agreement with their bank to remain open late (say 7 pm) and to clear the funds immediately, thus transferring the interest benefit sooner. Interest on a payment of $1 million is approximately $1,235 per day, or $3,000 over the two-day weekend and $4,300 for a long weekend. This is worth the effort required.

SUB-CONTRACTING VERSUS DIRECT LABOUR

There are many reasons why contractors have shifted from the traditional approaches to management (in-house resources) to outsourcing through sub-contract and supply agreements. It is generally held that contractors gain advantage in:

- Access to a greater range of labour and skill opportunities.
- Ability to manage fluctuations in workload by buying labour on an 'as required' basis.
- Ability to delegate specialised management tasks to sub-contract organisations
- Avoidance of employment laws relating to such matters as long service, holidays and termination.
- Flexibility in resourcing.
- Spreading the burden of financing the project.
- Allows under-capitalised firms to provide service.

Elazouni and Metwally (2000) suggest that 'subcontracting is a practice that contractors rely on to partially finance projects'. The ability to gain access to a cash flow stream, for indirect work, is another, related, aspect to the subcontracting issue. This must be a primary motivator supporting the contractor's desire to move toward subcontracting, and thus providing enormous leverage on their own corporate resources.

When the low capitalisation of construction firms is considered, and with the trend toward management only, it can be seen that for a very small investment there is a huge return. The funds generated from operations are potentially far greater than the value of the organisation managing them. This, properly managed, gives the contracting organisation significant opportunities, but leads also to significant risk for the system.

It is unlikely these days that a building firm will undertake direct labour works. The opportunity cost is too high, as they would be taking on board the funding role currently provided by trade credit. This would greatly limit the number and size of projects they could undertake.

SIGNS OF A DISTRESSED CONTRACTOR

There are several indicators of a construction contractor that is having trouble managing its organisational cash flow and that give clear warning to those doing business with that company. These are indicators only and there may be other reasons for these behaviours which have nothing to do with cash flow. These are not the same as balance sheet indicators of insolvency—such as may be undertaken by a credit rating agency, but rather indicators derived from observation of the behaviour of a company in the market place. Davis argues that many indicators of insolvency:

> are matters of commonsense but sometimes the same action could be interpreted both as prudent management and a distress signal (Davis, 1991).

Unilaterally extending the terms of payment

In this situation the contractor is likely to be confronting committed outward payments from past or current projects and is not generating sufficient surplus cash from current work. This might be due to a market downturn or delays in projects. Extending terms by, say, one extra month, does more than just put off the pain for a month. It has a multiplicative effect. Assuming thirty-day terms were normal before the change, the outward cash flow stream is made up of components, some of which are paid within the first month, some of which are paid in the second month (and sometimes this runs out into the third month as discussed in Chapter 6). With a change to sixty-day terms, there are now payments spread over the three months. This is illustrated in Figure 8.20.

The cross section suggest that this type of change has the potential to double the cash reserves being generated by extending terms from thirty days to sixty days. This will enable the contractor to remain in business and ride out short-term problems, but will do nothing to resolve long-term problems as payments still come due eventually.

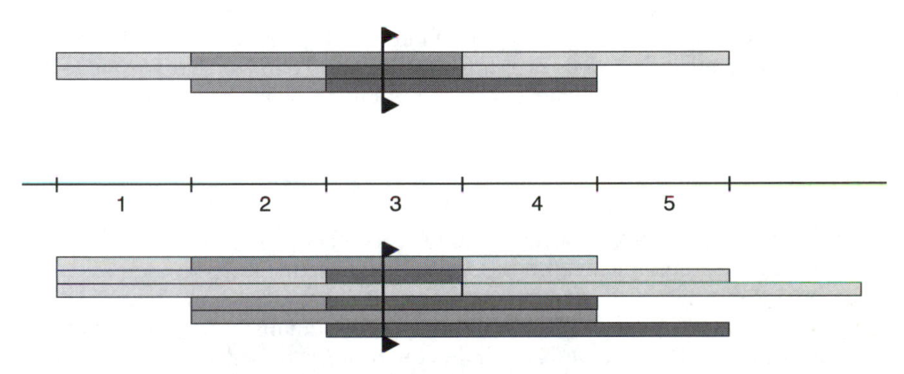

Figure 8.20 Extending payment terms

Requests for early payment

Requests for early payment may indicate a contractor with a short-term cash flow problem and may be an early warning of insolvency (Davis, 1991). However, strategic management of cash flow encourages early receipt of income, thus this could easily be a misleading indicator and would require further evidence in support.

Reduced quality of work

Firms that are struggling to maintain their cash flow will wish to reduce their costs. They may cut costs with the work, potentially leading to problems with quality of the finished work. This may be seen as a subtle indicator of the financial position of the firm—maybe not on its own, but in association with other factors.

Increased litigation

Davis (1991) argues that contractors may reschedule outgoings by raising disputes, or may introduce unjustified contra-charges to delay payment. They may also be more inclined to fight claims in preference to seeking a settlement, in an effort to squeeze the most out of the dispute or to further delay settlement (if payment is required).

Cost cutting may also contribute to some of these disputes. Contractors will be more likely to dispute a claim from a subcontractor or supplier or may try and withhold payment citing defects or similar concerns. Similarly, clients dissatisfied with the standard of their work will be inclined to withhold payment pending rectification of the work. Contractors will then pursue these outstanding payments.

The affect of these factors is that they will find themselves being subjected to, or initiating disputes. The resulting cost of legal action is sometimes blamed for the eventual failure of the company, as legal expenses can rapidly exceed the value of the issue being disputed, and while deferring a reckoning, ultimately payments must be made.

Increased disputation is an indicator of a financially distressed firm that prefers to pursue claims rather than settle disputes amicably, and one which is willing to burn subcontractors and clients to achieve a short-term increase in cash flow.

Staff turnover

An interesting research project would be to find out whether or not there is a relationship between the recent change of CEO and the subsequent insolvency of a firm. While looking like the subsequent incompetency of the new CEO, most people would recognise this as a 'hospital pass' to use a sporting colloquialism. Davis (1991) agrees that changes of staff can be an early indicator of a distressed firm. This turnover comes about through a number of factors, such as getting out while you still can (rats abandoning the sinking ship), or being driven out by additional demands through stretched resources or dissatisfaction with the work quality or environment.

Apparent success through new work

It seems strange to suggest that success is an indicator of failure, yet this is exactly the case. If it is accepted that maintenance of cash flow is paramount and that funds can be generated through projects, then it follows that in an organisational cash flow crisis, generating new work will temporarily solve the problem. The more work won, the better the apparent solution. A firm that suddenly wins more than its usual share of work is doing something different. It is possible this may be due to an innovative approach or management system, but in this industry it seems more likely that it comes through underbidding. Thus underbidding to win more work is an indicator of insolvency (Davis, 1991). This insolvency may come through the original problem not being solved, or through the compounding effect of work won using non-profitable margins, eventually making matters worse as projects complete.

SECURITY OF PAYMENTS

Liens, bonds, warrantees and insurance

The securities available to clients or contractors may play a strategic role in the payment system. These are important because they can provide protection in the system but at the same time they reduce the competitiveness of the sector. There is a natural desire to install safeguards and securities into the construction payment system to protect the players usually, if the rhetoric is to be believed, from defaulting clients. This desire is understandable, but it is important to address the interaction between freeing the system to encourage work and restricting the system to provide safety in the system.

It is clear, from the focus of this book on the payment system and the strategic management by contractors of their cash flow stream, that contractors' poor management strategies can result in flow on problems in the event of their failure which far exceed outstanding claims from errant clients. However, it is also true that many of those failures would have been brought to collapse by that single action of not paying that last claim. The system needs to consider both of those issues when considering security.

The system is also dependent on its competitiveness for success. For example, if contractors make it too onerous or expensive for clients to participate, then they will be discouraged from building or developing. Speculative developers would undertake only those projects with clearly viable margins. New businesses looking for a home would be discouraged, looking to existing stock for their solutions or perhaps going off-shore. To countries like Australia and New Zealand, where attracting new businesses is both necessary and difficult, the country cannot afford to erect barriers to new players.

It is useful to return to the underlying principle of the progress payment when resolving the issue of security. Are they a loan by the client to the contractor to allow them to undertake the work, or are they a payment for work completed and now in the possession of the client and therefore recoverable by the client in the event of a contractor failure? The answer is that, as 'title vanishes as goods are fixed to the land' (Davis, 1991) progress payments are somewhere between, which is why the industry has such complex systems and encounters so many difficulties.

Security in the form of a hold of assets is often touted as a solution. Providing a Lien over the title of the land until all claims are met would provide security for a contractor. The equivalent security for a client would be for the contractor to hand over the deeds to their company until all works are complete. The first discourages clients from entering into projects because at the end they may be prevented from making a rapid sale, being locked into disputes by a vexatious contractor, and therefore suffering loss. This is not in the best interests of the construction industry. The first is increasingly rare due to its complexity in operation and the way it discourages clients from investing, the second, while not actually being used in practice, would discourage contractors from participating in projects because they would consider the security far outweighed the risk. The two are, in function, very similar.

Davis (1991: 88) discusses the 'builders lien', stating:

> Contractors often speak of their right to possession of the site as a 'builders' lien'. Under English law, however, contractors do not have a lien over the site for prepaid monies and even jurisdictions conferring liens do not intend that they should entitle contractors to deny employers possession of the site. Rather, they impose a charge on the land or the employer's interest in it as security for the unpaid balance of the contract sum. The concept of a builders' lien over the site which does prevent the employer regaining possession does, however, appear in South Africa and possibly other jurisdictions.

For a detailed analysis of the workings of the 'Wages protection and Contractors' Liens act' in one such jurisdiction, refer to Wilson (1976), published before the repeal of that act in New Zealand.

Payment of all money at the conclusion of the project would protect the client, but would expose the contractor to the costs of funding the work (requiring the progress payment 'loan') and would expose them to the risk that the client could default without a Lien, holding no title to the completed works for which they had paid. Payment of all money at the start of the project would protect the contractor, but would expose the client to the risk of a contractor default and the loss of their prepayments, furthermore, partially completed work is of questionable value due to the reluctance of new contractors to take on incomplete structural work, the damage that may occur due to weather while settling disputes, etc.

The true risk to the client is the difference between the amount they have paid at any time and the value of the work in place, usually indicated by the cost to complete the work. The true risk to the contractor is the difference between the committed expenditure and the amount they have actually received from the client. Neither of these risks is sufficient to warrant draconian measures such as placing a Lien over the site or surrendering title to a company. Minimising securities to these risks is in the best interests of the industry, because it will encourage the most work.

The issue comes down to the timing of the payment. If the client pays in arrears as is the current practice, then the contractor requires additional security. If the client pays in advance then the client is the party requiring additional security. These securities could be provided in the form of bonds, warrantees or insurance.

The industry would be best served by a system that rewarded contractors with payment in advance, allowing them then the maximum flexibility for the strategic management of their cash flow. In return they would have to provide security of performance to the client and their ability to do so would be directly related to the strength and quality of their organisation. This would place well managed companies at a competitive advantage. Such a system would also remove the need for retention to be held against contractors, further freeing the system.

There are many other factors at work here which are beyond the scope of this discussion. It is, however, important to focus in the interaction between the freeing up and international competitiveness of the overall industry and the cost to society as a result of company failures which result from inadequate safeguards in the system.

For further information about the technical meaning of bonds and guarantees, refer to Davis (1991: Chapter 12) who places these terms within their legal framework. This highlights the differences between guarantees and bonds and the different types of bonds such as payment bonds and performance bonds.

Legislation

Construction industry legislation, commonly called security of payment legislation, is often held forward as a solution to the problems of corporate failure with flow-on damage to subcontractors. Such legislation generally provides for rapid resolution of disputes, with pay now resolve later approaches, quick mechanisms for bringing issues to a head rather than deferring until the impact snow-balls.

Such legislation does not generally proscribe payment terms (neither should it) but rather acts through prompt action in the event of dispute. Without determining payment terms, there is no consequence for strategic management of organisational cash flow as this operates within the terms of contracts (except for certain dubious practices) and is not restricted. As it acts through resolution of disputes only, such legislation does not address the quietly efficient subcontractor/contractor system which is generally so effective and efficient and which operates predominantly without rancour.

Legislation will only marginally impact on some of the negative consequences of cash farming, by bringing matters to a head sooner, and thereby reduce the damage.

Legislation which addresses the risks in the system, by addressing the protection of the gap in the payment system discussed above, provides more protection for the system without risking a reduction in competitiveness.

CONCLUSION

The strategic management of organisational cash flow is one of the most important issues for the long-term viability of a construction company. Evidence for the use of cash farming techniques in the building and construction industry has been found in the literature, and the evidence of company failures with the consequential flow-on affects is clear. The mechanisms of its operation and the conditions required for successful use have been proposed. In practice, cash farming is often a negative influence, however, the positive value of strategic management of cash flow to a well managed and funded organisation empowers a company to be more competitive. The organisational cash flow simulation undertaken here showed a clear improvement for an organisation able to separate the inflow and outflow profiles within the contractual relationships.

This knowledge has implications for change in the industry, and these should be considered prior to any change in policy with respect to the industry, such as the introduction of new legislation. Major clients, such as government, should also consider their role in funding the industry, and assess their procurement strategies.

These issues must necessarily be combined with research into the industry's supply chains. The subcontractors' and suppliers' roles in financing the industry should not be under estimated.

REFERENCES

Ashley, D. B. and Teicholz, P. M. (1977). 'Pre-estimate cash flow analysis'. *Journal of the Construction Division, American Society of Civil Engineers*, Proc. Paper 13213, **103** (C03): 369–379.

Ashman, G. B. (1994). *Security of Payments*. Melbourne, Victorian Government Publisher.

Cooke, B. and Jepson, W. B. (1979). *Cost and Financial Control for Construction Firms*. London, Macmillan.

Davis, R. (1991). *Construction insolvency*. London, Chancery Law Publishing.

Elazouni, A. M. and Metwally, F. G. (2000). 'D-SUB: Decision support system for subcontracting construction works' *Journal of Construction Engineering and Management*, **126**(3): 191–200.

Gyles, R. V. (1992). *Royal Commission into Productivity, Report of the Hearings*. Part 1. Sydney, State Government of New South Wales.

Kaka, A. P. (1996). 'Towards more flexible and accurate cash flow forecasting'. *Construction Management and Economics* **14**: 35–44.

Kaka, A. P. and Price, A. D. F. (1991). 'Net cashflow models: Are they reliable'? *Construction Management and Economics* **9**: 291–308.

Khosrowshahi, F. (2000). 'A radical approach to risk in project financial management', in *ARCOM Sixteenth annual conference*, Glasgow. Glasgow Caledonian University, (2) 547-556.

Nazem, S. M. (1968). 'Planning contractor's capital'. Building Technology and Management October: 256–260.

Punwani, A. (1997) 'A study of the growth-investment-financing nexus of the major UK construction groups'. *Construction Management and Economics*, **15:** 349–361

Tong, Y. and Lu Y. (1992). 'Unbalanced bidding on contracts with variation trends in client-provided quantities'. *Construction Management and Economics* **10**: 69-80.

Wilson, J. N. (1976). *Contractors' Liens and Charges*, 2nd edition. Wellington, Butterworths.

Authors Index

E

F

G

H

I

J

K

L

M

N

P

R

S

T

V

W

Y

Z

General index

A

B

C

D

E

F

G

I

K

L

M

N

O

P

R

S

T

U

W